Sustainable Textiles: Production, Processing, Manufacturing & Chemistry

Series Editor

Subramanian Senthilkannan Muthu, Chief Sustainability Officer
Green Story, Canada
Tung Chung, Hong Kong

This series aims to address all issues related to sustainability through the lifecycles of textiles from manufacturing to consumer behavior through sustainable disposal. Potential topics include but are not limited to: Environmental Footprints of Textile manufacturing; Environmental Life Cycle Assessment of Textile production; Environmental impact models of Textiles and Clothing Supply Chain; Clothing Supply Chain Sustainability; Carbon, energy and water footprints of textile products and in the clothing manufacturing chain; Functional life and reusability of textile products; Biodegradable textile products and the assessment of biodegradability; Waste management in textile industry; Pollution abatement in textile sector; Recycled textile materials and the evaluation of recycling; Consumer behavior in Sustainable Textiles; Eco-design in Clothing & Apparels; Sustainable polymers & fibers in Textiles; Sustainable waste water treatments in Textile manufacturing; Sustainable Textile Chemicals in Textile manufacturing. Innovative fibres, processes, methods and technologies for Sustainable textiles; Development of sustainable, eco-friendly textile products and processes; Environmental standards for textile industry; Modelling of environmental impacts of textile products; Green Chemistry, clean technology and their applications to textiles and clothing sector; Eco-production of Apparels, Energy and Water Efficient textiles. Sustainable Smart textiles & polymers, Sustainable Nano fibers and Textiles; Sustainable Innovations in Textile Chemistry & Manufacturing; Circular Economy, Advances in Sustainable Textiles Manufacturing; Sustainable Luxury & Craftsmanship; Zero Waste Textiles.

Subramanian Senthilkannan Muthu
Editor

Corporate Social Responsibility in Textiles and Fashion

Springer

Editor
Subramanian Senthilkannan Muthu
Chief Sustainability Officer
Green Story, Canada
Tung Chung, Hong Kong

ISSN 2662-7108 ISSN 2662-7116 (electronic)
Sustainable Textiles: Production, Processing, Manufacturing & Chemistry
ISBN 978-3-031-61098-1 ISBN 978-3-031-61099-8 (eBook)
https://doi.org/10.1007/978-3-031-61099-8

This Springer imprint is published by the registered company Springer Nature Switzerland AG
The registered company address is: Gewerbestrasse 11, 6330 Cham, Switzerland

This book is dedicated to:
The lotus feet of my beloved Lord
Pazhaniandavar
My beloved late Father
My beloved Mother
My beloved Wife Karpagam and Daughters
—Anu and Karthika
My beloved Brother
Last but not least
To everyone working in the textile sector to
make it "SUSTAINABLE"

Contents

Natural Coloring Agents in Raising Awareness of Indonesian Textile Industry Sustainability and the Role of Corporate Social Responsibility: Health and Environmental Aspects

Anastasia Wheni Indrianingsih and Cici Darsih

Abstract The textile industry is very important for human life because clothing and textiles are one of the basic human needs besides food and shelter. Humans need clothes to maintain their survival, from hot and cold weather conditions and for personal protection. However, with the development of technology and fashion, besides the main purposes above, clothing and textiles have also become an art for humans. One form of art here is using various dyes in the textile industry. Unfortunately, these synthetic dyes negatively affect the environment and living creatures, including humans. It has spurred the return of textile coloring to nature, namely by using natural dyes that are environmentally friendly and safer for health. In this chapter, emphasis is given on how corporate social responsibility (CSR) can play a very important role in influencing textile SME industries in Indonesia to encourage the use of natural dyes in their business processes, besides also discussing possible sources of natural coloring agents, how to extract natural coloring agents and the advantages of natural coloring for the environment and the health of living creatures.

Keywords Corporate social responsibility · Dyes extraction · Functional clothing textile · Natural coloring agent

1 Introduction

One of the biggest influences on business industries in the world is the textile industry, which has a major impact on the state's economy (Pertiwi et al., 2020). This is supported by the fact that many countries are using their textile industry to expand

A. W. Indrianingsih (✉) · C. Darsih
Natural Products Research Group, Research Center for Food Technology and Processing, National Research and Innovation Agency (BRIN), Yogyakarta, Indonesia

S. S. Muthu (ed.), *Corporate Social Responsibility in Textiles and Fashion*, Sustainable Textiles: Production, Processing, Manufacturing & Chemistry, https://doi.org/10.1007/978-3-031-61099-8_1

their exports. It opens up opportunities and generates significant revenue for the country. As in Indonesia, textile and garment export growth increased throughout 2019 with a value of US$ 13.8 billion (https://www.aseanbriefing.com, October 2023). Even though several industries have been affected by the COVID-19 outbreak and the US-China trade war, the textile industry in Indonesia includes large industry and SME industries (Small-Medium Enterprises). Batik is one kind of textile industry in Indonesia, generally established in the Java region. Pekalongan is the region where the batik industry has the most impact on Indonesia's economy. There are 1342 small industries in Pekalongan, and 83.1% are batik industries (Fajri, 2013). Those industries produce 300–1000 pieces of cloth monthly (Kusumastuti et al., 2022). The textile industry not only has an impact on the economy but also has an impact on the environment and society. One form of industrial responsibility to positively impact the environment and society is through CSR programs.

Corporate Social Responsibility (CSR) is generally interpreted as a commitment from a business company to contribute to society's economics near the workplace (World Business Council for Sustainable Development, 2000). CSR programs aim to improve society's life, especially the local community. Another literature (Blowfield & Frynas, 2005) states that the meaning of CSR is that companies should have a relationship with their stakeholders with a responsibility for social and environmental issues. Meanwhile, a previous study revealed that in the CSR concept, several responsibilities must be considered important aspects, including legal, economic, ethical, and philanthropic (Carroll, 1991). Legal aspects are related to the law, how to ensure that all business processes comply with applicable legal regulations; economic aspects are related to how to obtain benefits that the local community can utilize; ethical aspects are related to the decision or company policy taken must be by the generally applicable code of ethics; and the philanthropic aspect means that the company must run its company well, not be dangerous, be fair, and be a good corporation. The other concept of CSR states that economic, legal, ethical, and philanthropic aspects are in an equally important position, and they interconnect with each other (Schwartz & Carroll, 2003). However, this concept and definitions of CSR may vary from organization to organization, company to company, country location and time (Gray, 2010).

Therefore, this study aims to provide an overview of the understanding of Indonesian textile companies, including Batik as traditional clothing companies, which are widely spread in Indonesia, both large and small scale (small and medium enterprises, SMEs) regarding engagement in CSR practices, to find out what factors encourage CSR practices, factors that inhibit CSR practices, and in general, the role of CSR in the textile industry in Indonesia in the fields of health and the environment. Based on existing approaches, in this chapter, we use several literatures related to the topic by Scopus and Google Scholar search and based on the experience carried out by the authors during technology transfer related to natural colors in batik cloth SMEs in Indonesia. In Sect. 2, we propose the source of natural agents that are widely available in nature, while the extraction method to obtain natural coloring agents is presented in Sect. 3. In Sects. 4 and 5, we discuss about the application of natural coloring agents in the Indonesian textile industry and the role of CSR in this business. Finally, the conclusion is provided in Sect. 6.

2 Sources of Natural Coloring Agent

Natural coloring agents, such as pigment and dyes, are commonly used in medicine, food, and textiles, which are one of the important needs for human life (Wainwright, 2008; Ghosh et al., 2022). Even though pigments and dyes have the same functions as coloring agents, they are different in their properties, such as dyes having higher solubility than pigments. The utilization of pigments and dyes began long years ago, as reported that ancient Romans, Egyptians, Greeks, and Indians used them as natural coloring agents (Abel, 2021). However, the success of producing synthetic dyes has reduced the use of natural coloring agents. Synthetic dyes have several advantages, such as low cost, a wide variety of color ranges, easy application, and high durability (Hagan & Poulin, 2021). Synthetic dyes are produced in around 800,000 t worldwide to meet the textile industry's needs (Slama et al., 2021). Despite their advantages, synthetic coloring agents have bad environmental and health effects, such as reducing water quality, inhibiting plant growth, impairing photosynthesis, bioaccumulation, and increasing carcinogenicity and mutagenicity (Lellis et al., 2019).

Meanwhile, only about 1% of natural coloring agents are being used compared to synthetic dyes due to the poor coloring ability of natural coloring agents and the high cost of natural coloring agent production (Krizova, 2015). Natural resources can provide natural coloring agents using sustainable extraction methods to fulfill the industries' needs with good performance. It will be important to achieve a better impact on environmental and human health.

The source of natural coloring agents can come from plant parts such as fruit, bark, wood, flowers, roots, leaves, and seeds. The coloring from plants can also be a combination of plant parts. Another source of natural coloring agents can also be animals and minerals (Sigurdson et al., 2017). The use of minerals as natural colorants was observed many years ago in Africa and Europe. The classification of natural coloring agents consists of inorganic and organic agents, in which inorganic colorant consists of pigments while organic colorant consists of pigment and dyes (Bhattacharyya, 2010). Inorganic pigments are commonly made from minerals such as cadmium (red, orange, and yellow), titanium (white), and cobalt (blue). Natural colorants from inorganic pigments are obtained from the ground.

Meanwhile, organic pigments have many more advantages besides being in greater quantity than inorganic pigments, including low toxicity, purity in the shade and excellent color (Pandian et al., 2021). The low toxicity of organic pigments was because of their immobility properties and low solubility in organic solvents and water. On the other hand, most natural dyes obtained from organic sources can be classified based on their chemical structure as carotenoids (Mortensen, 2006), flavonoids (Khan et al., 2011), indigoids (Islam et al., 2022), betalains (Ali &

El-Mohamedy, 2011), unionists (Khan et al., 2012), pyridine based (Leona & Lombardi, 2007), dihydropyran based (Nilesh et al., 2015) and tannins (Shabbir et al., 2017).

On the other hand, Indonesia has many types of traditional cloth, such as songket cloth, batik cloth, tapis cloth, gringsing cloth, and ikat cloth (Rahayu et al., 2020). This clothing originally used natural dyes. More than 40 plants have been used as natural dyes in textile dyeing in Indonesia (Indraningsih & Darsih, 2013). However, as time passes, synthetic dyes are being used more widely because of the above-mentioned advantages. The CSR program of the textile industry in Indonesia can play an important role in reusing natural coloring agents by providing education to local SMEs. That will increase public awareness, especially in local communities, about the importance of using natural coloring agents in the traditional textile and fabric industry, especially Batik, for human health and environmental sustainability.

3 Extraction of Natural Dyes

Natural dyes are classified into four classes based on chemical structures, sources, color produced, and application method. The main step in preparing natural dyes is the extraction method. The determination of the extraction method can affect the yield and purity of natural dyes. The extraction method chosen must also be appropriate according to the solubility of the pigment or dyes (Gupta, 2019; Pizzicato et al., 2023). There are different kinds of extraction methods to obtain dyes.

(a) Aqueous Extraction

This method is suitable for extracting dyes that are soluble in water. Generally, the source is cut into small pieces and macerated using water after the boiling process for one night. The dyes obtained from this method can be directly used to color fibers, natural yarn, or textiles. The advantages of this method are safe technique, sustainability, and ease. In contrast, the disadvantages of this method are a small amount of yield, long extraction time, required large amount of water, decomposition of dyes that are sensitive to temperature, and many impurities such as sugar and other water-soluble compounds (Gupta, 2019; Pizzicato et al., 2023; Saxena & Raja, 2014). Berhanu and Ratnapandian (2017) reported that the optimum condition for dye extraction from the bark of *Cassia singueana* plan was achieved when the bark was extracted using water with a concentration of 60 g/L at 95 °C for 60 min. This dye can be used for dyeing sheepskin crust leather with a natural mordant such as aloe vera juice. Another study reported that the maximum conditions for extracting mango leaves using water were at pH 10, a temperature of 98 °C for 60 min, with a material-to-liquor ratio of 1:10 (Uddin, 2015). In Indonesia, small textile industries such as the Batik Industry also generally use this method.

(b) Solvent Extraction

Natural dyes can be extracted using polar solvents and nonpolar solvents such as hexane, chloroform, petroleum ether, ethanol, methanol, a mixture of water and alcohol, and a mixture of ethanol with methanol. The limitation of this method is that it requires specific instruments, toxic residues, greenhouse gases, and other unwanted compounds extracted together with natural dyes, such as chlorophyll and wax. Meanwhile, the solvent method obtains higher yield, lower temperatures, easy to remove and reuse the solvent, and low degradation of dyes (Pizzicato et al., 2023). Previous studies showed the extraction of Mangifera indica and Caesalpinia sappan bark using maceration, Soxhlet, and reflux techniques with ethanol and methanol solvents. The results revealed that the optimum ratio of material to solvent in the extraction process was 1:6, and reflux followed by maceration was suitable for the textile industry (Sutrisna et al., 2020).

(c) Acid and Alkali Extraction

The acid and alkali extraction methods are similar techniques to the previous method. The difference is that acid or alkali extraction is used to extract glycoside dyes from natural sources. Alkali and acid solutions hydrolyze the glycoside chain in dyes. Alkali extraction is suitable for extracting dyes containing phenolic compounds. This method's extraction obtained a higher yield, but some dyes are sensitive to pH and could be degraded (Bahlool et al., 2023; Pizzicato et al., 2023; Salauddin Sk et al., 2021). *Cocos nucifera* husk has been extracted using NaOH with various concentrations and extraction times. The dyes obtained in this study were then applied to cotton fabric using mordant tannic acid. The research reported that the concentration of NaOH and extraction time influenced the yield of dyes, and 7.5% tannic acid was the best concentration of mordant for improving cotton's fastness properties (Rodiah et al., 2021).

(d) Ultrasonic and Microwave-Assisted Extraction

Ultrasonic and Microwave-Assisted Extraction are modern techniques used to extract dyes from natural resources. In Ultrasonic-Assisted Extraction (UAE), acoustic cavitation disrupts the cell material. Pressure alteration in ultrasonics is caused by transferring the energy of ultrasonics into the liquid solution. The breaking of cells by intense shock waves, shear force, microbubble resonance, local high temperature, and pressure are the main mechanical/physical in ultrasonic extraction (Liu et al., 2022). Several researchers have used this method to extract natural dyes (Ma et al., 2019; Qadariyah et al., 2018; Sadeghi-Kiakhani et al., 2019). Natural dye from Hawthorn fruits was extracted using the UAE technique with various solvents such as methanol, acetone, ethanol, water, and a mixture of ethanol and water. The higher yield of this method was obtained at a sample weight of 8 g/L, at pH 5, 60 °C for 40 min. The natural dyes obtained in this study were then applied to Polyamide (nylon 6) fabric, and the result showed high colorfastness properties and antibacterial activity (Sadeghi-Kiakhani et al., 2021). Meanwhile, the main mechanism of

Microwave-Assisted Extraction MAE) is the generation of heating and dipole rotation in organic molecules in the sample. The frequency and wavelength in microwaves usually range between 0.3 and 300 GHz and 0.001 and 1 m. Both methods, UAE and MAE, have the same advantages, such as low energy, temperature, solvents, high yield, and eco-friendly; these methods do not need complicated instruments and are inexpensive (Abo-State et al., 2013; Hamdy et al., 2021; Pizzicato et al., 2023; Ramola et al., 2019; Salauddin Sk et al., 2021). Wizi et al. reported that the Ultrasound-microwave-assisted (UMA) extraction method using a mixture of solvent ethanol and water (7:3) with HCl significantly improved the extraction of colorants from sorghum husk (Wizi et al., 2018).

(e) Enzymatic Extraction and Fermentation

The enzymatic and fermentation method is an eco-friendly extraction process for natural dyes. This method avoids using solvents in the extraction process but instead uses enzymes to extract active compounds from the plant matrix (Pizzicato et al., 2023). An example of this method is the fermentation of indigo dye from the Indigofera plant (Laitonjam & Wangkheirakpam, 2011; Li et al., 2019).

(f) Supercritical Fluids

The Supercritical fluids method is harmless, clean, safe, cheap, and widely available, with no residual solvent, no environmental pollution, less time for extraction, good selectivity, and low temperature (Borges et al., 2012; Muthu, 2016). However, this method is the most complex process for extracting natural dyes. *Acacia cyanophylla* flowers, *Tamarix aphylla* leaves, tomato processing residues, and Pepper (*Capsicum annuum* L.) by-products were extracted using supercritical carbon dioxide (99.9% CO_2). The results showed that supercritical CO_2 improves fiber's dyeability (Dhouibi et al., 2021).

4 Use of Natural Dyes in the Textile Industry (Including Batik) in Indonesia

Increasing utilization of natural dyes in the textile industry because of the awareness of environment and health. The advantages of natural dyes compared to synthetic ones are renewable resources, biodegradable, eco-friendly, exhibit functional properties, easy extraction and purification, unique shade of color, and the waste and product can be used as sources (Che & Yang, 2022; Gupta, 2019). The natural dyes can be extracted from renewable sources such as plants, animals, and microorganisms (Nambela et al., 2020). Natural dyes are biodegradable and eco-friendly because they are easy to decompose and nontoxic in the environment. Some research reported that natural dyes possess functional properties in the textile, such as antibacterial, anti-UV, antioxidant, deodorizing, anti-creasing, and mothproof (Che & Yang, 2022). Pomegranate peel and onion peel have been extracted and used for dyeing the cotton yard (Botteri et al., 2022). The disadvantages of natural dyes are poor affinity and uptake on textile, require mordants, reproducibility of shades, no standard color recipes and methods available, average washing fastness, no standard extraction method, and mostly applicable for natural fibers (Gupta, 2019; Pizzicato et al., 2023). The natural dyes depend on the weather, geographical conditions, species and maturity of sources, and time of harvesting that influence the shade of dyes. The limited shade range, for example, the blue color of natural dye, is only obtained from indigo sources (Saxena & Raja, 2014). Some natural dyes need mordant to improve their affinity, fastness properties, and final shades on the fibers.

The activities of natural dyes on human health have been reported in many literatures. Several natural dyes such as anthocyanins, betalains, curcumins, and carotenoids have shown several biological activities such as anti-inflammatory activity, antioxidant activity, anti-cardiovascular activity, anticancer activity, antidiabetic activity, antiviral activity, anti-obesity, neuroprotective effects, and antimicrobial activity (Li et al., 2022). Carotenoids can scavenge reactive oxygen species in the human body, thus having good antioxidant activity (Dose et al., 2016). Anthocyanins from red clover (*Trifolium pratense*) have shown anti-inflammatory effects by inhibiting the gene expression such as interleukin-1 beta (IL-1β), tumor necrosis factor-alpha (TNF-α), cyclooxygenase-2 (COX-2), and monocyte chemoattractant protein (MCP-1) (Lee et al., 2020). In anticancer activity, natural dyes such as β-carotene, lycopene, fucoxanthin, and astaxanthin have been suggested to inhibit various cancers, including breast cancer cells (Ahn et al., 2020), while anthocyanin has been reported to inhibit lung cancer, colorectal cancer, prostate cancer, breast cancer, and liver cancer (Charepalli et al., 2015). In the antidiabetic activity, astaxanthin can decrease blood glucose levels dose-dependently (Fen et al., 2021). Anthocyanins also show anti-obesity activity by inhibiting lipid absorption, suppressing food intake, and regulating lipid metabolism (Xie et al., 2018). Antiviral activity has been shown by delphinidin, one kind of anthocyanin, by altering the particle structure of the hepatitis C virus (Calland et al., 2015). Natural dyes such as curcumin have been found to prevent skin from UV light (Yusuf et al., 2017), while natural dyes extracted from orange peel have been found to have UV protection than synthetic dyes (Shahid et al., 2013). Antibacterial clothing was also developed by incorporating *Garcinia mangostana* extract (Indrianingsih et al., 2021). Hopefully, antimicrobial textiles will increase consumer health because using antibacterial cloth will protect the skin from possible infections and bacterial attacks. Furthermore, using natural coloring agents will guarantee the safety and health of consumers and workers in the textile industries since natural coloring agents are generally non-dangerous substances.

Several studies have reported the beneficial effects of natural dyes on the environment. Berberine, lawsone, and curcumin are dye molecules that perform corrosion inhibition against several metals due to their structure, which consists of N, S, and O atoms (Johnsirani et al., 2015). The biodegradable and renewable materials of natural dyes also positively impact the environment. An example of a renewable material is farmers planting Indigofera tinctoria plants (Elsahida et al., 2019), which could maintain the sustainability of a natural coloring agent. Since natural coloring agents are generally safe for human health, other materials used in textile coloring, called mordants, can also contaminate the environment since they usually use chemicals such as aluminum sulfate (Rinawati et al., 2017). Thus, biomordants from plant extracts can solve this problem and reduce the possibility of environmental pollution. Some of the biomordants that have already been studied are pomegranate rind (*Punica granatum*) (Deshmukh and Wath, 2011) and *Ixora javanica* leaves (Indrianingsih et al., 2021).

5 The Role of CSR in the Textile Industry in Indonesia

Several aspects are motivating factors for CSR activities, such as ethical commitment, philanthropy, improving the image of the company, instructions from top management, improving the company performance, and attention from society (Pintol & Allui, 2020). In Indonesia, the possibility of CSR activities by companies is based on the company's commitment to advancing the surrounding community's economic sector. Also, by carrying out CSR activities, the company can become well-known by the public and the government with a positive image, which will positively influence their business processes. Similarly, it has been revealed that in Saudi Arabia, companies have CSR activities to improve their reputation in the market. Thus, they are trustworthy companies for managing risk (Zahari et al., 2020).

The obstacle faced in the CSR program for natural dyes in the Indonesian textile industry is the commitment of all parties to implement the program. Some textile industries still use synthetic dyes for the Batik dyeing process because they think the process is easier and cheaper than natural dyes. The skills, personalities, and limitations on the amount of natural dyes in the industry also hinder the CSR Program. These challenges were similar to CSR in the Arabian Organization (Pintol & Allui, 2020).

The CSR programs can provide benefits for women who live in industrial environments and for the industry itself. These women can work in industry so they can increase their family income. Apart from that, these women can also improve their skills and knowledge outside their main job as housewives and gain confidence (Maertens & Swinnen, 2009). Meanwhile, the benefits of CSR programs for the industry itself include increased business performance, new marketing, increased and rising profits in the industry, and increased customer confidence in the industry (Szegedi et al., 2023; Thorisdottir & Johannsdottir, 2020).

A previous study revealed that CSR activities have many advantages for the local community and the company. It can also positively affect their stakeholders in a positive way (Yin, 2017). Several advantages to companies that conduct CSR activities include developing a company's positive image, inviting new customers, raising profit, greater brand and reputation, increasing employee relations, increasing innovation and productivity, and stronger relations with communities (Asemah, 2013). Moreover, a company's CSR activities are a strategic act to impact society positively while doing business to increase its profit. Thus, companies need to satisfy society and obtain good performance on financial in return.

6 Conclusion

The present study provides an overview of the CSR activities of natural coloring agents in the Indonesian textile industry. However, CSR programs related to the application of natural dyes in the textile industry in Indonesia are still not running well because many factors still need to be strengthened and developed. The textile industry has many challenges related to many sectors, such as politics, economics, environment, and health. CSR activities in natural dyeing in the textile industry will benefit the surrounding community, especially in health and environmental fields, and improve the company's positive image. Textile industry players' awareness of the use of natural coloring agents must be promoted to maintain sustainability in all sectors. Furthermore, a commitment from all parties, including government, companies, consumers, researchers, and society, is needed to create a healthy and sustainable climate in the textile industry, especially in using natural coloring agents.

References

Abel, A. (2021). The history of dyes and pigments: From natural dyes to high performance pigments. In *Colour design* (pp. 557–587). DCC Europe Ltd. https://doi.org/10.1016/B978-0-08-101270-3.00024-2

Abo-State, M. A., Ragab, A. M. E., El-Gendy, N. S., Farahat, L. A., & Madian, H. R. (2013). Effect of different pretreatments on egyptian sugar-cane bagasse saccharification and bioethanol production. *Egyptian Journal of Petroleum, 22*(1), 161–167. https://doi.org/10.1016/j.ejpe.2012.09.007

Ahn, Y. T., Kim, M. S., Kim, Y. S., & An, W. G. (2020). Astaxanthin reduces stemness markers in bt20 and t47d breast cancer stem cells by inhibiting expression of pontin and mutant P53. *Marine Drugs, 18*, 577.

Ali, N. F., & El-Mohamedy, R. S. R. (2011). Eco-friendly and protective natural dye from red prickly pear (*Opuntia lasiacantha* Pfeiffer) plant. *Journal of Saudi Chemical Society, 15*(3), 257–261. https://doi.org/10.1016/j.jscs.2010.10.001

Asemah, E. S. (2013). Business advantages of corporate social responsibility practice: A critical review. *New Media and Mass Communication, 18*, 45–54.

Bahlool, S. O., Kenawy, Z. M., El-Baset, A., & Yasser, A. (2023). Preparation of absorbent medical cotton from non-spinnable Egyptian cotton fibers. *Journal of Textiles, Coloration and Polymer Science, 20*(2), 137–145.

Berhanu, T., & Ratnapandian, S. (2017). Extraction and optimization of natural dye from Hambo Hambo (*Cassia singueana*) plant used for coloration of tanned leather materials. *Advances in Materials Science and Engineering, 2017*, 7516409. https://doi.org/10.1155/2017/7516409

Bhattacharyya, N. (2010). *Natural dyes for textiles and their eco-friendly applications* (p. 22). IAFL Publications.

Blowfield, M., & Frynas, J. (2005). Editorial setting new agendas: Critical perspectives. Corporate social responsibility in the developing world. *International Affairs, 81*, 499–513.

Borges, M. E., Tejera, R. L., Díaz, L., Esparza, P., & Ibáñez, E. (2012). Natural dyes extraction from cochineal (*Dactylopius coccus*). New extraction methods. *Food Chemistry, 132*(4), 1855–1860.

Botteri, L., Miljković, A., & Glogar, M. I. (2022). Influence of cotton pre-treatment on dyeing with onion and pomegranate peel extracts. *Molecules, 27*(14), 4547.

Calland, N., Sahuc, M. E., Belouzard, S., Pene, V., Bonnafous, P., Mesalam, A. A., Deloison, G., Descamps, V., Sahpaz, S., Wychowski, C., et al. (2015). Polyphenols inhibit hepatitis C virus entry by a new mechanism of action. *Journal of Virology, 89*, 10053–10063.

Carroll, A. (1991). The pyramid of corporate social responsibility: Toward the moral management of organizational stakeholders. *Business Horizons, 34*(4), 39–48.

Charepalli, V., Reddivari, L., Radhakrishnan, S., Vadde, R., Agarwal, R., & Vanamala, J. K. P. (2015). Anthocyanin-containing purple-fleshed potatoes suppress colon tumorigenesis via elimination of colon cancer stem cells. *The Journal of Nutritional Biochemistry, 26*, 1641–1649.

Che, J., & Yang, X. (2022). A recent (2009–2021) perspective on sustainable color and textile coloration using natural plant resources. *Heliyon, 8*(10), e10979. https://doi.org/10.1016/j.heliyon.2022.e10979

Deshmukh, A., & Wath, M. (2011). Pomegranate rind as a natural mordant in dyeing of cotton with *Carissa karandas* Linn. *Journal of Bionano Filter, 4*(2), 299–301.

Dhouibi, N., Baaka, N., Charradi, R., Bouine, I., Dhaouadi, H., & Dridi-dhaouadi, S. (2021). Multi-fiber dyeing improvement using natural supercritical CO_2 extracts. *Fibers and Polymers, 22*(7), 1874–1882. https://doi.org/10.1007/s12221-021-0422-2

Dose, J., Matsugo, S., Yokokawa, H., Koshida, Y., Okazaki, S., Seidel, U., Eggersdorfer, M., Rimbach, G., & Esatbeyoglu, T. (2016). Free, radical scavenging and cellular antioxidant properties of astaxanthin. *International Journal of Molecular Sciences, 17*, 103.

Elsahida, K., Fauzi, A. M., Sailah, I., & Siregar, I. Z. (2019). Sustainability of the use of natural dyes in the textile industry. *IOP Conference Series: Earth and Environmental Science, 399*, 012065.

Fajri, P. Y. N. (2013). *Spatial modeling for determining the location of the wastewater treatment plant of batik industry in Pekalongan, Central Java*. Master Dissertation, Institut Pertanian Bogor, Bogor, Indonesia.

Fen, Z. G., Ni, Y. H., Wan, C. Y., Liu, F., & Fu, Z. W. (2021). Anti-diabetic effects of astaxanthin on an stz-induced diabetic model in rats. *Endocrine Journal, 68*, 451–459.

Ghosh, S., Sarkar, T., Das, A., & Chakraborty, R. (2022). Natural colorants from plant pigments and their encapsulation: An emerging window for the food industry. *LWT, 153*, 112527. https://doi.org/10.1016/j.lwt.2021.112527

Gray, R. (2010). Is accounting for sustainability actually accounting for sustainability…and how would we know? An exploration of narratives of organizations and the planet. *Accounting, Organizations and Society, 35*, 47–62.

Gupta, V. K. (2019). *Fundamentals of natural dyes and its application on textile substrates* (A. K. Samanta, N. S. Awwad, & H. M. Algarni, Eds., Chapter 2). IntechOpen. https://doi.org/10.5772/intechopen.89964

Hagan, E., & Poulin, J. (2021). Statistics of the early synthetic dye industry. *Heritage Science, 9*, 33. https://doi.org/10.1186/s40494-021-00493-5

Hamdy, D., Hassabo, A. G., & Othman, H. A. (2021). Various natural dyes using plant palette in coloration of natural fabrics. *Journal of Textiles, Coloration and Polymer Science, 18*(2), 121–141.

Indraningsih, A. W., & Darsih, C. (2013). Natural dyes from plants extract and its applications in indonesian textile small medium scale enterprise. *Eksergi Journal, 11*(1), 16–22. https://doi.org/10.31315/e.v11i1.327

Indrianingsih, A. W., Rosyida, V. T., Darsih, C., & Apriyana, W. (2021). Antibacterial activity of Garcinia mangostana peel-dyed cotton fabrics using synthetic and natural mordants. *Sustainable Chemistry and Pharmacy, 21*, 100440. https://doi.org/10.1016/j.scp.2021.100440

Islam, M. T., Farhan, M. S., Faiza, F., Halim, A. F. M. F., & Sharmin, A. A. (2022). Pigment coloration research published in the science citation index expanded from 1990 to 2020: A systematic review and bibliometric analysis. *Colorants, 1*, 38–57. https://doi.org/10.3390/colorants1010005

Johnsirani, V., Rajendran, S., Mary, C. A., Catherine, M., Joseph, R., Umasankareswari, T., & Sundari, J. J. (2015). Corrosion inhibition by dyes from plants. *International Journal of Nano Corrosion Science and Engineering, 2*, 22–28.

Khan, M. I., Ahmad, A., Khan, S. A., Yusuf, M., Shahid, M., Manzoor, N., & Mohammad, F. (2011). Assessment of antimicrobial activity of Catechu and its dyed substrate. *Journal of Cleaner Production, 19*(12), 1385–1394. https://doi.org/10.1016/j.jclepro.2011.03.013

Khan, S. A., Ahmad, A., Khan, M. I., Yusuf, M., Shahid, M., Manzoor, N., & Faqeer Mohammad, F. (2012). Antimicrobial activity of wool yarn dyed with *Rheum emodi* L. (Indian Rhubarb). *Dyes and Pigments, 95*(2), 206–214. https://doi.org/10.1016/j.dyepig.2012.04.010

Krizova, H. (2015). Natural dyes: Their past, present, future and sustainability. In *Recent developments in fibrous material science* (pp. 59–71). https://dspace.tul.cz/handle/15240/29724

Kusumastuti, A., Atika, Achmadi, T. A., Phusavat, K., & Hidayanto, A. N. (2022). Assessment of producer's perspective on the production of environmentally friendly fashion products: A case study in Indonesian natural dyes batik craftsmen. *Environmental Science and Pollution Research, Fajri, 30*, 124767. https://doi.org/10.1007/s11356-022-23330-z

Laitonjam, W. S., & Wangkheirakpam, S. D. (2011). Comparative study of the major components of the indigo dye obtained from *Strobilanthes flaccidifolius* Nees. and *Indigofera tinctoria* Linn. *International Journal of Plant Physiology and Biochemistry, 3*(7), 108–116.

Lee, S. G., Brownmiller, C. R., & Lee, S. (2020). Anti-inflammatory and antioxidant effects of anthocyanins of trifolium pratense (red clover) in lipopolysaccharide-stimulated RAW-267.4 macrophages. *Nutrients, 12*, 1089.

Lellis, B., Fávaro-Polonio, C. Z., Pamphile, J. A., & Polonio, J. C. (2019). Effects of textile dyes on health and the environment and bioremediation potential of living organisms. *Biotechnology Research and Innovation, 3*(2), 275–290. https://doi.org/10.1016/j.biori.2019.09.001

Leona, M., & Lombardi, J. R. (2007). Identification of berberine in ancient and historical textiles by surface-enhanced Raman scattering. *Spectroscopy, 38*(7), 853–858.

Li, S., Cunningham, A. B., Fan, R., & Wang, Y. (2019). Identity blues: The ethnobotany of the indigo dyeing by Landian Yao (Iu Mien) in Yunnan, Southwest China. *Journal of Ethnobiology and Ethnomedicine, 15*(1), 1–14.

Li, N., Wang, Q., Zhou, J., Li, S., Liu, J., & Chen, H. (2022). Insight into the progress on natural dyes: Sources, structural features, health effects, challenges, and potential. *Molecules, 27*, 3291. https://doi.org/10.3390/molecules27103291

Liu, Y., Liu, X., Cui, Y., & Yuan, W. (2022). Ultrasound for microalgal cell disruption and product extraction: A review. *Ultrasonics Sonochemistry, 87*, 106054. https://doi.org/10.1016/j.ultsonch.2022.106054

Ma, X., Wei, Y., Wang, S., Zuo, X., & Shen, B. (2019). Sustainable ultrasound-assisted ultralow liquor ratio dyeing of cotton fabric with natural turmeric dye. *Textile Research Journal, 90*(5–6), 685–694. https://doi.org/10.1177/0040517519878793

Maertens, M., & Swinnen, J. (2009, March 31–April 2). *Are African high-value horticulture supply chains bearers of gender inequality*. Exposé à l'occasion de l'atelier FAO–Fonds International de Dévelopement Agricole– OIT Sur Le Thème "Gaps, trends and current research in gender dimensions of agricultural and rural employment: Differentiated pathways out of poverty". Rome.

Mortensen, A. (2006). Carotenoids and other pigments as natural colorants. *Pure and Applied Chemistry, 78*(8), 1477–1491. https://doi.org/10.1351/pac200678081477

Muthu, S. S. (2016). *Textiles and clothing sustainability: Recycled and upcycled textiles and fashion* (p. 1). Springer. https://doi.org/10.1007/978-981-10-2146-6

Nambela, L., Haule, L. V., & Mgani, Q. (2020). A review on source, chemistry, green synthesis and application of textile colorants. *Journal of Cleaner Production, 246*, 119036. https://doi.org/10.1016/j.jclepro.2019.119036

Nilesh, P., Mithun, N. S., Rajput, S., Rangabhatla, G. S., Prasad, V., & Ahmad, M. (2015). Brazilin from Caesalpinia sappan heartwood and its pharmacological activities: A review. *Asian Pacific Journal of Tropical Medicine, 8*(6), 421–430. https://doi.org/10.1016/j.apjtm.2015.05.014

Pandian, B., Ramalingam, S., Sreeram, K. J., & Rao, J. R. (2021). Natural pigment: Preparation of brown pigment from lignin biomass for coloring application. *Dyes and Pigments, 195*, 109704. https://doi.org/10.1016/j.dyepig.2021.109704

Pertiwi, C. D., Irawati, A., & Chen, J. F. (2020). The effect of corporate social responsibility disclosure on financial performance: Comparative study of textile companies listed on Indonesia and Taiwan stock exchanges. In *Proceeding International Conference on Information Technology and Business* (pp. 17–25).

Pintol, L., & Allui, A. (2020). Critical drivers and barriers of corporate social responsibility in Saudi Arabia organizations. *Journal of Asian Finance, Economics and Business, 7*(11), 259–268. https://doi.org/10.13106/jafeb.2020.vol7.no11.259

Pizzicato, B., Pacifico, S., Cayuela, D., Mijas, G., & Riba-Moliner, M. (2023). Advancements in sustainable natural dyes for textile applications: A review. *Molecules, 28*(16), 1–22. https://doi.org/10.3390/molecules28165954

Qadariyah, L., Mahfud, M., Sulistiawati, E., & Swastika, P. (2018). Natural dye extraction from teak leves (tectona grandis) using ultrasound assisted extraction method for dyeing on cotton fabric. *MATEC Web of Conferences, 156*, 4–7. https://doi.org/10.1051/matecconf/201815605004

Rahayu, M., Kuncari, E. S., Rustiami, H., & Dewi. (2020). Utilization of plants as dyes and natural color binder in traditional pringgasela woven fabric, east lombok, West Nusa Tenggara, Indonesia. *Biodiversitas, 21*(2), 636–641. https://doi.org/10.13057/biodiv/d210228

Ramola, B., Kumar, V., Nanda, M., Mishra, Y., Tyagi, T., Gupta, A., & Sharma, N. (2019). Evaluation, comparison of different solvent extraction, cell disruption methods and hydrothermal liquefaction of Oedogonium macroalgae for biofuel production. *Biotechnology Reports, 22*, e00340. https://doi.org/10.1016/j.btre.2019.e00340

Rinawati, D. I., Sari, D. P., Purwanggono, B., & Hermawan, A. T. (2017). Environmental impact analysis of batik natural dyes using life cycle assessment. *AIP Conference Proceedings, 1902*, 020044. https://doi.org/10.1063/1.5010661

Rodiah, M. H., Noor Hafizah, S., Noor Asiah, H., Nurhafizah, I., Norakma, M. N., & Norazlina, I. (2021). Extraction of natural dye from the mesocarp and exocarp of *Cocos nucifera*, textile dyeing, and colour fastness properties. *Materials Today: Proceedings, 48*, 790–795. https://doi.org/10.1016/j.matpr.2021.02.315

Sadeghi-Kiakhani, M., Hashemi, E., & Gharanjig, K. (2019). Inorganic nanoparticles and natural dyes for production of antimicrobial and antioxidant wool fiber. *3 Biotech, 9*(12), 1–16. https://doi.org/10.1007/s13205-019-1974-3

Sadeghi-Kiakhani, M., Tehrani-Bagha, A. R., Safapour, S., Eshaghloo-Galugahi, S., & Etezad, S. M. (2021). Ultrasound-assisted extraction of natural dyes from Hawthorn fruits for dyeing polyamide fabric and study its fastness, antimicrobial, and antioxidant properties. *Environment, Development and Sustainability, 23*, 9163–9180.

Salauddin Sk, M., Rony, M. I. A., Haque, M. A., & Shamim, A. M. (2021). Review on extraction and application of natural dyes. *Textile and Leather Review, 4*(4), 218–233. https://doi.org/10.31881/TLR.2021.09

Saxena, S., & Raja, A. S. M. (2014). Natural dyes: Sources, chemistry, application and sustainability issues. In S. S. Muthu (Ed.), *Roadmap to sustainable textiles and clothing: Eco-friendly raw materials, technologies, and processing methods* (pp. 37–80). Springer. https://doi.org/10.1007/978-981-287-065-0_2

Schwartz, M., & Carroll, A. (2003). Corporate social responsibility: A three-domain approach business. *Ethics Quarterly, 13*(4), 503–530.

Shabbir, M., Islam, S. U., Bukhari, M. N., Rather, L. J., Khan, M. A., & Mohammad, F. (2017). Application of *Terminalia chebula* natural dye on wool fiber – Evaluation of color and fastness properties. *Textiles and Clothing Sustainability, 2*(1), 1–9. https://doi.org/10.1186/s40689-016-0011-8

Shahid, M., Islam, S. U., & Mohammad, F. (2013). Recent advancement in natural dye applications: A review. *Journal of Cleaner Production, 53*, 310–331. https://doi.org/10.1016/j.jclepro.2013.03.031

Sigurdson, G. T., Tang, P., & Giusti, M. M. (2017). Natural colorants: Food colorants from natural sources. *Annual Review of Food Science and Technology, 8*, 261–280. https://doi.org/10.1146/annurev-food-030216-025923

Slama, H. B., Chenari, A., Bouket, Z., et al. (2021). Diversity of synthetic dyes from textile industries, discharge impacts and treatment methods. *Applied Sciences, 11*, 6255. https://doi.org/10.3390/app11114625

Sutrisna, P. D., Hadi, R. P., Valentina, J. J., Priyantini, H. R., Waluyo, P. W., & Ronyastra, I. M. (2020). Natural dyes extraction intended for coloring process in fashion indutries. *IOP Conference Series: Materials Science and Engineering, 833*(1), 012093. https://doi.org/10.1088/1757-899X/833/1/012093

Szegedi, K., Németh, T., & Körtvési, D. (2023). Employer branding in the fashion industry: CSR actions by fashion SMEs. *Sustainability (Switzerland), 15*(3), 1–18. https://doi.org/10.3390/su15031827

Thorisdottir, T. S., & Johannsdottir, L. (2020). Corporate social responsibility influencing sustainability within the fashion industry. A systematic review. *Sustainability (Switzerland), 12*(21), 1–64. https://doi.org/10.3390/su12219167

Uddin, M. G. (2015). Extraction of eco-friendly natural dyes from mango leaves and their application on silk fabric. *Textiles and Clothing Sustainability, 1*(1), 1–8. https://doi.org/10.1186/s40689-015-0007-9

Wainwright, M. (2008). Dyes in the development of drugs and pharmaceuticals. *Dyes and Pigments, 76*(3), 582–589. https://doi.org/10.1016/j.dyepig.2007.01.015

Wizi, J., Wang, L., Hou, X., Tao, Y., Ma, B., & Yang, Y. (2018). Ultrasound-microwave assisted extraction of natural colorants from sorghum husk with different solvents. *Industrial Crops and Products, 120*, 203–213.

World Business Council for Sustainable Development. (2000). *Corporate social responsibility: Making good business sense*. Retrieved from http://ceads.org.ar/downloads/Making%20good%20business%20sense.pdf. Accessed 05 Oct 2023.

Xie, L., Su, H., Sun, C., Zheng, X., & Chen, W. (2018). Recent advances in understanding the anti-obesity activity of anthocyanins and their biosynthesis in microorganisms. *Trends in Food Science and Technology, 72*, 13–24.

Yin, J. (2017). Institutional drivers for corporate social responsibility in an emerging economy: A mixed-method study of Chinese business executives. *Business & Society, 56*, 672–704.

Yusuf, M., Shabbir, M., & Mohammad, F. (2017). Natural colorants: Historical, processing and sustainable prospects. *Natural Products and Bioprospecting, 7*, 123–145. https://doi.org/10.1007/s13659-017-0119-9

Zahari, A. R., Esa, E., Rajadurai, J., Azizan, N., & Muhamad Tamyez, O. F. (2020). The effect of corporate social responsibility practices on brand equity: An examination of Malaysia's top 100 brands. *Journal of Asian Finance, Economics and Business, 7*(2), 271–280.

Downstream Corporate Social Responsibility

Use-Oriented Business Models in Scandinavian Apparel Industry

Cecilia Mark-Herbert and Moa Nyvall

Abstract Corporate social responsibility (CSR) entails addressing sustainability challenges in the entire value system, from sourcing, producing, to downstream CSR that involves the use of a product. This chapter focuses on a service-dominant logic corporate perspective in outdoor apparel business models in Scandinavia. The aim of this chapter was to identify enabling factors for use-oriented business models. A collective case study, analyzing marketing relationships, was made in the business model Canvas. Offering services, in this case rental, in addition to traditional sales puts demand on getting to know the customers' needs and preferences. Other enabling factors were also related to additional services, other activities in support of the rental, resources such as IT solutions, and communication serving educational needs.

Keywords Added value · Concept · Comparative case study · Outdoor apparel · Rental · Service dominant logic · Textile

1 Introduction

The industrial revolution (late 1700 to mid 1800) paved the way for economic growth through efficient manufacturing processes, automation, and economy of scale production (Kónya & Ohashi, 2004). Focus in this development was placed on material and financial wealth. These objectives are reflected in business models that were designed to maximize outcomes, using a linear take-make-dispose logic that motivated companies to maximize profit by selling high quantities of products that were manufactured at a minimal cost (EMF, 2013). However, with our current

C. Mark-Herbert (✉) · M. Nyvall
Department of Forest Economics, Swedish University of Agricultural Sciences, Uppsala, Sweden
e-mail: Cecilia.mark-herbert@slu.se; Moa.nyvall@atrax.se

S. S. Muthu (ed.), *Corporate Social Responsibility in Textiles and Fashion*, Sustainable Textiles: Production, Processing, Manufacturing & Chemistry,
https://doi.org/10.1007/978-3-031-61099-8_2

understanding of sustainable development (El Alfy et al., 2020; Moro et al., 2022), it is clear that the product-focused business model induces unsustainable use of natural resources, increased consumption rates, and high waste generation, all of which represent problems in industries all over the world.

A product-focused institutional context is reflected in legal frameworks that support unsustainable consumption patterns. In the case of textiles (cloths and home textiles[1]) in Sweden, the market increased by 40% in just a decade, from 2000 to 2009 (Carlsson et al., 2011, p. 17). The reason for the significant increase is assumed to be a combination of growing prosperity and lower textile prices. At the same time, 8 kilos of textiles per person are disposed of in Sweden annually (Tojo, 2012, p. 44), which is almost as high as the European annual average of textile disposal, 11 kilos per person (EC, 2022). Global textile production almost doubled between 2000 and 2015. There are a number of sustainability challenges related to the textile system in the world (Muthu, 2017), such as resource depletion and global warming (Schor, 2010; Shuk, 2016).

This chapter opens with a review of business models as a precondition for downstream, use-oriented corporate responsibility strategies. A case study of outdoor apparel service providers in Sweden, based on interviews and secondary empirical material, is analyzed using the Canvas business model as a starting point. In the following analysis and discussion, conditions for product-service system models are compared to what is recognized in academic literature as parts of a business model. Identified empirical factors in this case are captured in a discussion of factors that influence the development of use-oriented business models in terms of needed resources, partners, and activities. The contributions in this chapter are captured in understandings of resource efficiency, where consumers contribute to reduced material flow, reverse logistics, and cultural adoption factors.

1.1 Needs for New Business Models

Business models are traditionally designed according to a linear take-make-dispose logic, which motivates companies to maximize profit by selling large quantities of products that are manufactured at minimal costs (EMF, 2013). However, the need for business models that shift focus from selling products to providing services, enabling firms to create value with less resource usage, has been recognized for centuries as part of corporate social responsibility (Charter & Polonsky, 1999; Bocken et al. 2014; McCoy & Chi, 2022). These models can potentially decouple the satisfaction of consumer needs from environmental impacts (Tunn et al., 2021). It assumes multiple material loops of reusing, repairing, and remanufacturing a product before it turns to waste (EMF, 2013). The phenomenon of making more use of products through services is referred to as product-service systems (Baines et al.,

[1] In this chapter, we use the terms textile, apparel, and cloths as synonyms.

2007; Mont, 2004; Tukker, 2015; Tukker & Tischner, 2006). This circular system entails a mix of products and services that together fulfill customer needs. Businesses profit by maximizing units-of-services delivered.

1.2 Problem: Needs for a New Business Model

Corporate Social Responsibility (CSR) is a concept that has gained significant prominence in the business world over the past few decades. It refers to a company's commitment to integrating ethical, social, and environmental concerns into its business operations and interactions with stakeholders. CSR goes beyond the traditional focus on profit maximization and emphasizes the importance of contributing positively to society and the environment. CSR initiatives are expressed in a variety of corporate choices related to, for example, sourcing, employee well-being, community engagement, and philanthropy. In some cases, CSR activities assume stakeholder engagement, whether in upstream ethical sourcing procedures or downstream use and disposal of products by consumers. This chapter focuses on *downstream activities* in a use-oriented product-service system, where the business model changes the value proposition for its customers. It refers to activities that engage consumers in resource management, in this case related to the use of apparel.

Use-oriented product-service systems are based on principles of collaborative consumption, such as leasing, renting, or pooling (Tukker, 2004), none of which are novel phenomena. For some product categories, this is the norm. An example is outdoor associations that commonly offer rentals of ski equipment (Klingelhöfer, 2018) or bikes (Ahn et al., 2022). However, renting out a wide range of consumer goods for casual use on a large scale is a new trend (Åkerlund, 2018; Batat, 2019; Eidse, 2022; Mark-Herbert & Byfors, 2023). Start-ups are emerging that offer rentals and subscription-based leasing services (Myin et al., 2022). New actors that take the lead with innovative service offerings (Witell et al., 2016) may pressure established manufacturing companies to follow the rental trend in order to stay competitive (Bessant & Davies, 2007). Already established manufacturing companies that currently sell products may, however, find it challenging to shift focus to services (Kindström, 2010). Bocken et al. (2014) advocate that it requires fundamental changes in the core business, its purpose, and daily processes to alter a business model.

Due to the challenges in terms of organizational changes required, there are few examples of a product manufacturer that has fully transitioned into a service provider (Oliva & Kallenberg, 2003). Rarely, if ever, is an example presented of a company that previously sold products and has succeeded in shifting to merely renting out or leasing products. There are not sufficient guidelines and tools for practitioners to transition from a product to a service-dominant logic business model (Baines et al., 2007). Integrating services into an existing product offer requires changes to the organizational structure, management, and culture (Barquet et al.,

2013; Kindström, 2010). If these elements are not considered carefully, businesses risk falling into "the service paradox," which entails that the firm has invested heavily in service offerings, the operations require high costs and no returns are realized (Gebauer et al., 2005). Reim et al. (2015) argue that companies seeking to make the shift towards services must engage in an organizational transformation in the way they create, deliver, and capture value.

From a theoretical perspective, most business models assume that profit is made on sales of a product, and ownership by the consumer after the transaction is taken for granted (Bocken et al., 2014; Lewandowski, 2016). This is referred to as product-dominant logic. Little attention is given to service aspects in the business model that enable *other* ownership structures of products. Meier et al. (2010) argue that more investigation is needed on how businesses may capture values that are co-created with consumers meanwhile retaining ownership of a product. This calls for alternative views of meeting consumer needs, such as what the product-service system epistemology offers.

1.3 Aim, Research Questions, and Delimitations

Developing a business by altering the business model to meet sustainability objectives can be seen as a corporate responsibility strategy (Mark-Herbert & Prejer, 2018). It entails developing a business model that addresses sustainability challenges in the entire value system with a wide stakeholder definition. In this chapter, we explain how the Scandinavian outdoor retail sector is making efforts to transform its business models. The aim of the project on which this chapter is based was to identify enabling factors for use-oriented business models. Research questions of particular interest relate to the characteristics of a use-oriented business model and managing perceived challenges in the Scandinavian outdoor apparel industry:

- What are the characteristics of a use-oriented business model?
- What are the perceived challenges in introducing use-oriented services in combination with the existing product offer?

Delimitations in the project are made in theoretical and empirical dimensions. The concept of a product-service system is commonly used in engineering, computer science, and business management (Baines et al., 2007; Boehm & Thomas, 2013; Tukker, 2015). This project focuses on business management and marketing perspectives with a particular interest in business model theory. The Business Model Canvas was selected as a conceptual framework because it is easily understood and adopted by practitioners. It is also a suitable structure for explaining product-service systems. Empirical delimitations are made in terms of data from one point in time, which does not allow for an analysis of an organizational change process. These delimitations support the descriptive contribution of this chapter.

2 Approach

The approach in this chapter is presented in two dimensions, in terms of methodological (Sects. 2.1, 2.2, 2.3, and 2.4) and theoretical (Sects. 2.5 and 2.6) choices. The approach is characterized by a flexible research design of a case study in the apparel industry. Initially, a literature review was carried out to identify the research forefront, focusing on business models and use-oriented services.

2.1 *Method*

A case study enables the investigation of a complex phenomenon in its real-life context (Eisenhardt, 1989; Yin, 2013). The case study approach helps set boundaries for the study object and specify the area of interest (Creswell, 2013). In this study, the area of interest is the use-oriented business model within the Scandinavian outdoor apparel industry. The reason this study focuses on the outdoor apparel industry is twofold. Firstly, a majority of case studies in existing product-service system literature are conducted within the heavy industries of tools and chemicals (Reim et al., 2015). Few studies explore the concept within more easily accessible consumer goods where buying decisions are taken more frequently (Mont et al., 2006). To enable a wider diffusion of product-service systems, it is interesting to investigate products that are consumed on a daily basis by a wider range of consumers, such as clothes and other outdoor equipment. Secondly, products within the outdoor apparel industry are generally durable and of high quality, which is suitable when introducing a use-oriented service since it intensifies product use.

A multiple case study of use-oriented services in outdoor equipment in Scandinavia serves as the empirical base for this chapter. As argued by Yin (2013), a multiple case study is preferred since it enables identifying similarities and differences between the cases. This case study is of a descriptive nature since it aims to describe the use-oriented business model in its specific context.

2.2 *Case Corporations as One Unit of Analysis*

Case companies were selected according to three criteria: (i) having business offers rental- and/or leasing services and (ii) business operating in the Scandinavian outdoor apparel industry. To assure inclusion of multiple views from actors in different positions in the value chain, both manufacturers and retailers were contacted. Seven companies agreed to participate in this case study: Fjällräven, Houdini, Bergans, Naturkompaniet, Rent-a-plagg, Aktivt Uteliv, and Re:leased. Since investigating business models requires a close investigation of how companies create, deliver and capture values (Kindström, 2010; Osterwalder & Pigneur, 2010), the unit of analysis in this study is the firm.

2.3 Data Collection, Analysis, and Quality Assurance

A pilot study was carried out in an initial corporate document analysis and interview to test the conceptual perspectives in interview questions in preparation for the case representatives (Gunnarsson, 2019). In-depth interviews were conducted in this project to gain detailed descriptions of the meanings and perceptions related to the use-oriented business model (Table 1). When using a case study approach, in-depth interviews are an essential method to provide details and rich descriptions (Yin, 2013). As proposed by Kvale (1996), the interviews in this project were well prepared, and the interviewer had a fundamental understanding of the specific topic in order to ensure internal validity. Secondly, there is a risk of receiving biased answers from respondents who want to provide a glorified image of their company. Respondents were therefore informed that the aim of this study is *not* to compare different companies but rather to get an overall picture of the use-oriented business model.

Interviews were structured with the support of themes from the Canvas model themes in open-ended questions. Having open-ended questions allows the researcher to adapt the conversation according to the respondent's answers and ask additional questions about specific issues (Kvale, 1996). This was particularly valuable in this study since the theoretical framework was constantly evolving during the data collection process. Yin (2013) advocates that a wider understanding can be obtained of the phenomenon if conducting multiple interviews with more than one person.

In this project, a mix of personal interviews and phone interviews was conducted for practical reasons. The dialogue in each interview was opened with a presentation of the project, the respondent's role, and the procedures, as part of informed consent following the guidelines by the university (VR, 2017). The phone interviews enabled increased privacy when discussing sensitive topics as well as a smaller interviewer effect (Novick, 2008). In addition, three personal interviews were conducted face-to-face, enabling observation and visual cues of the context, facial expressions, and body language. The personal interviews were held at respondents' offices, and

Table 1 Data collection in the case study (validated interviews, in 2019)

Role and company	Type of interview	Interviewee
Project Manager Circular business models, Fjällräven	Personal	Elsa
Sustainability Manager, Fjällräven	Personal	Denise
Business Developer, Fjällräven	Personal	Pelle
CEO, Its Re:leased	Phone	Jonna
CEO, Fjällräven	Phone	Mikael
CEO, Aktivt Uteliv	Phone	Edvin
Sustainability Developer, Bergans	Phone	Anna
	Phone	Hans
CEO, Rent a plagg	Phone	Markus
Retail Manager, Houdini	Email	Patrik
Business Developer, Houdini	Seminar	Robin

the familiar environment led to a relaxed atmosphere. Due to time constraints, one of the respondents answered the interview questions by email. After the interviews were conducted, additional questions were asked via email if clarification was needed. All interviews by phone and face-to-face were recorded, transcribed, and sent to the respondent for validation.

2.4 Data Analysis and Quality Assurance of the Research Process

A content analysis, carried out in three steps as proposed by Elo et al. (2014), was used in the analysis of the empirical material. The three steps—preparation, organization, and reporting—may appear linear, but the process should be seen as iterative. In the preparation step, the pilot study serves as a test of the interview guide. Interviews were carried out parallel to writing up analytical texts that were gradually refined in line with what Miles and Huberman (1994) refer to as an inductive research process. The organization step refers to transcribing and coding the interview material in an iterative process. The last step in the analysis is that of reporting, which is part of the analysis process.

Grounds for quality assurance are created throughout the research process. Credibility, being confident that the findings are true and valid (Lincoln & Guba, 1985), is seen in this project through triangulation and validation of interview transcripts. Transferability refers to findings being applicable in other contexts (*ibid.*) and is closely related to generalization. Case studies cannot be statistically generalizable to a large population. That is not their role. However, they can provide grounds for making a transfer given contextual understandings.

The choice of analysis technique bears certain consequences. For instance, it is difficult to analyze a change process through content analysis since it is a rather static approach that highlights the content at a specific point in time. However, this study does not strive to explain the transition from products to services with a longitudinal approach. Instead, business model theory is used as a theoretical framework, which is also fixed in nature and illustrates a business logic at one point in time (Osterwalder & Pigneur, 2010). For this purpose, content analysis is desired in this study to withdraw common characteristics of the use-oriented business model.

2.5 A Theoretical Perspective

CSR perspectives in this chapter are presented in terms of downstream activities. These activities may relate to enabling consumers to be part of the efforts made to take responsibility. Empirical examples are found in information to guide consumers in their choices of what to buy related to the waste hierarchy or nudging consumers towards sustainability practices (Mont, 2004, Mont et al., 2006). In this chapter, we present a product portfolio strategy, *Product-Service System* that relates to offering services as an alternative to purchasing a physical product.

From a corporate perspective, Product-Service System (PSS) refers to value creation that combines the sale of products with related services to create value for customers (Blüher et al., 2020). Instead of selling a physical product as a standalone item, companies offering PSS provide a bundle of products and services to meet customer needs more comprehensively. This approach aims to enhance customer satisfaction, reduce environmental impacts, and create long-term relationships with customers (Barquet et al., 2013).

2.6 Business Model Adaptation to Product-Service System

A business model is a strategic framework that outlines how a company intends to create, deliver, and capture value in the market. It describes the fundamental aspects of how a business operates, generates revenue, and sustains itself. A well-defined business model is essential for guiding the company's activities, attracting investors, and ensuring long-term success. Figure 1 illustrates the well-established Canvas model (Osterwalder & Pigneur, 2010). It offers an overview of what is considered as key components of a business model: value proposition, target customer segments, revenue streams, cost structures, channels, key resources, activities, and customer relationships.

The Canvas model (Fig. 1) was developed from a product-dominant perspective, with physical products as the basis for value creation (Osterwalder & Pigneur, 2010). A *use-oriented business model*, on the other hand, refers to a strategic approach that emphasizes a specific orientation or focus in the way a company conducts its operations, serves its customers, and generates value from a consumer use

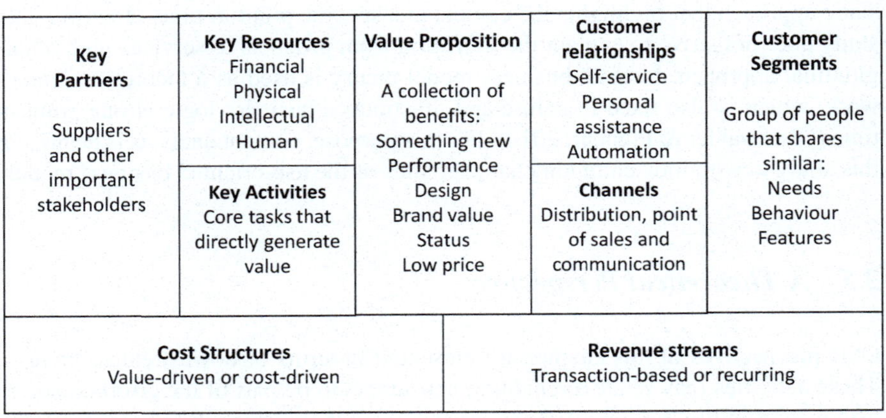

Fig. 1 Canvas business model, the nine components with some examples. (Based on Osterwalder & Pigneur, 2010, p. 40)

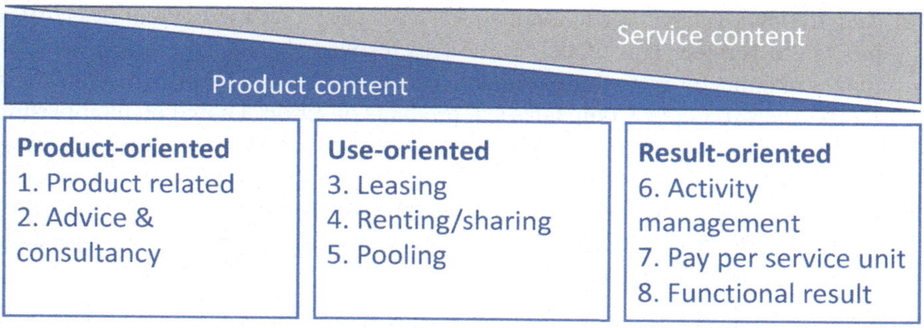

Fig. 2 Eight types of product-service system business models. (Based on Tukker, 2004, p. 248)

perspective. This orientation can be centered on various aspects of the business, such as customer-centricity, innovation, sustainability, or technology. It means focusing on how consumer needs are satisfied with a mix of products and services.

An example of a use-oriented business model is captured in the concept of Product-Service System. It is a business model that integrates the sale of products and services to satisfy customer needs and deliver value. Instead of simply selling a physical product, companies that adopt a PSS approach offer a combination of products and services that provide a more holistic solution to customers. This concept is often associated with sustainability and the shift towards a more circular economy. Key characteristics of different types of product-service systems are described by Tukker (2004) in Fig. 2.

Tukker's (2004) eight types of PSS business models present innovative approaches to delivering products and services that can reduce resource consumption and environmental impact (Fig. 2). These models promote sustainability by encouraging shared use, extended product life, and responsible consumption. The eight types of PSS business models as identified by Tukker (2004) are further presented below:

1. *Product-related:* In this model, the focus is on extending the lifespan of products through maintenance, repair, and refurbishment services. It can reduce the need for new products and minimize waste.
2. *Advice and consultancy:* Instead of selling products, companies offer them as services. Customers pay for the utility or functionality of the product without owning it, which can encourage manufacturers to design longer-lasting and more sustainable products.
3. *Leasing:* Companies pool resources and share them among multiple customers. For example, industrial equipment or appliances can be shared among several businesses, reducing the overall number of units needed.
4. *Renting or sharing*: This model facilitates the sharing of products or resources among users, such as car-sharing platforms, tool libraries, and co-working spaces. It promotes collaborative consumption and resource optimization.
5. Pooling: Customers subscribe to access a product or service, pooling, for a specific period, rather than purchasing it outright.

6. *Activity management*: Customers are charged based on their actual usage of a product or service. This model encourages responsible consumption and incentivizes manufacturers to design durable products.
7. *Pay per performance*: Companies sell products or services based on their performance or outcomes, rather than on the product itself. Manufacturers are motivated to ensure high-quality and efficient performance to meet contract terms.
8. *Functional unit*: This model allows customers to customize products according to their needs and preferences. It reduces overproduction and waste by producing only what is demanded.

These PSS business models are designed to align economic interests with sustainability goals, reducing resource consumption and environmental impact. They promote a circular economy by emphasizing resource efficiency, waste reduction, and the responsible use of products and services.

3 Results and Analysis

With the selected conceptual frameworks for value creation (the PSS and Canvas model), the case study of downstream CSR is further presented in terms of a brief literature review of use-oriented business models, an overview of previous research, followed by a presentation of the empirical case study.

3.1 Use-Oriented Business Models: Previous Studies

Previous studies of use-oriented (UO) business models (BM) in various areas of fast-moving consumer goods point to desired effects, such as resource efficiency (Hirschl et al., 2003; Ceschin, 2013), prolonged life cycles of products (Tu & Ngo, 2018), and low access barriers for new customers (Tukker, 2004). However, previous studies also point to risks for rebound effects where products are used less carefully, resulting in faster wear and tear (Tukker, 2004, 2015; Kuo, 2011; Reim et al., 2015) and accessibility expectations (Armstrong et al., 2015). An overview of some examples is provided in Table 2.

Bocken et al. (2014) highlight that the level of environmental performance when delivering functionality rather than ownership depends on the organization's ability to reduce production volume and eliminate waste. During a two-year project, Mont et al. (2006) studied the lease of baby prams in Sweden, and they argue that the positive environmental effects are determined by three criteria. Firstly, products should not be rented out through a third party, but instead directly by the manufacturer who should be interested in prolonging the product's life cycle. Secondly, once the product has been used intensively, it should not be sold but instead remain with the producer to be remanufactured or recycled. Thirdly, companies should operate

Table 2 Previous product system services (PSS) studies of use-oriented business models (UO BM)

Focus of investigation	Main findings	Author & year
PSS characteristics, benefits and barriers	Few examples of complete PSSs, companies lack a system approach	Mont (2002)
Environmental effects of rentals	Ski rentals increase resource productivity by 1.6	Hirschl et al. (2003)
Eight archetypes for PSSs BMs	Rental is inconvenient but low access barriers	Tukker (2004)
Lease baby prams	High initial costs but high long-term profit	Mont et al. (2006)
Designing PSS systems	Consumers handling uncertainties	Rexfelt and Ornäs (2009), Williams (2007)
A leasing simulation model	Leasing can result in rebound effects	Kuo (2011)
A literature review on PSS BMs and a single case study on machine lease	Long-term revenues and internal resistance are challenging. Suggest launching UO services under a separate business unit	Barquet et al. (2013)
Manufacturing companies' transition to PSSs	Successful transition requires service innovation and new communication approach	Parida et al. (2014), Armstrong et al. (2015)
Literature review of implementation of PSS	UO models require marketing ownerless consumption and new financial partners	Reim et al. (2015)
Possibilities to introduce rental services into the fashion industry	Rental requires changes in six of nine elements in canvas BM	Karlsson and Fernandes (2016)
Balance scorecards to evaluate servitization and identify improvement factors	Leasing implies intensified use of resources, new marketing operations, financial partners, profitability strategy and new service design	Tu and Ngo (2018)
PSS development for home furnishing	Barriers for developing PSS: financial and economic concerns, product design, capabilities, relationships, end-users, and policy	Schoonover et al. (2021)
Fashion rental platforms	Sustainability information needs to be very clear, cleaning rental items is key for fashion rental services	McEachern et al. (2020), McCoy and Chi (2022)

according to the circular waste hierarchy: reduce, reuse, and recycle. In a previous study, Mont (2002) emphasizes that companies commonly implement some parts, but there are few examples of a complete PSS that is designed to consider the whole life cycle of products. The reason is that firms often lack a system approach and PSSs have not been studied enough to assist in their practical application.

Bocken et al. (2014) highlight that the level of environmental performance when delivering functionality rather than ownership depends on the organization's ability to reduce production volume and eliminate waste. During a two-year project, Mont et al. (2006) studied the lease of baby prams in Sweden, arguing that the positive environmental effects are explained by three criteria. Firstly, products should not be rented out through a third party but instead directly by the manufacturer, who should be interested in prolonging the product's life cycle. Secondly, once the product has been used intensively, it should not be sold but instead remain with the producer to be remanufactured or recycled. Thirdly, companies should operate according to the circular waste hierarchy: reduce, reuse, and recycle. In a previous study, Mont (2002) emphasizes that companies commonly implement some parts, but there are few examples of a complete PSS that is designed to consider the whole life cycle of products. The reason is that firms often lack a system approach, and PSSs have not been studied enough to assist its practical application.

Karlsson and Fernandes (2016) investigated the possibilities of renting out clothes within the Swedish fast fashion industry through a case study and the theoretical lens of BM canvas. Their study indicates that the value proposition should shift focus from selling products to selling access, which will inevitably change the revenue streams. The study suggests that a web shop is a promising distribution channel that could contribute to competitive advantages since rentals have not been offered digitally in the Swedish fast fashion industry previously. Moreover, customers may be perceived as suppliers since they return the clothes to the producer, and it is crucial to treat them as partners. Several studies conclude that long-term customer relationships should be created through co-creation activities, and employees should be educated in sustainable consumption (Barquet et al., 2013; Reim et al., 2015; Karlsson & Fernandes, 2016). Moreover, research indicates that UO BMs imply new marketing operations to promote ownerless consumption (Barquet et al., 2013; Reim et al., 2015; Tukker, 2015; Tu & Ngo, 2018). Reim et al. (2015) explain that employees have a key role in promoting the PSS and selling the idea. It is not sufficient to merely list the benefits as when selling a product, but customers need to be able to visualize the advantages with the service. Therefore, Parida et al. (2014), McEachern et al. (2020), and McCoy and Chi (2022) point to the importance of communicating the added value with UO services through demonstrators and calculators.

Karlsson and Fernandes (2016) argue that the customer segment, key resources, and cost structures are the same in a rental model as when selling products. However, their conceptual framework merely departs from BM theory and does not consider literature on PSS. Another study indicates that all elements should change in a UO model except for the customer segment (Barquet et al., 2013). Mont et al. (2006) emphasize that UO PSS is characterized by long-term revenues, and the cost structures differ, therefore, substantially from PO models. Their study found that the cash flow is negative in the initial stage of leasing (the first 10 months in the baby pram case). To finance the added costs during the first years of implementing a UO model, studies suggest collaborating with a financial actor (Mont et al., 2006; Reim et al., 2015). However, a case study conducted by Tu and Ngo (2018) shows that a firm's products may not be pledged as loan collateral, which implies that financial actors may not be willing to enter into a partnership.

Despite a negative cash flow initially, a financial analysis conducted by Mont et al. (2006) and 15 years later by Schoonover et al. (2021) indicates that leasing generates higher revenues long-term per product compared to traditional sales. Even so, several studies indicate that employees are unaware of the potential value with UO offers and do not tend to believe in its profitability (Oliva & Kallenberg, 2003; Barquet et al., 2013; Tu & Ngo, 2018). Therefore, there may be internal resistance to introducing rental services as acknowledged by Barquet et al. (2013). One of the most significant barriers for companies going from product-focused to service-focused is the cultural change, which requires substantial time and resources (Mont, 2002). For instance, Oliva and Kallenberg (2003) provide an example of engineers who previously got paid full price for selling a qualitative product that they spent a lot of time and resources on developing. They may not be very excited about the firm receiving a tenth of that only for cleaning and repairing it.

Tukker (2015) concludes that managers are commonly afraid to initiate a transition towards services since it requires a large investment that is currently spent on manufacturing and product development, which is the firm's traditional source of gaining a competitive advantage. One way for the management team to signal the shift in focus to the rest of the organization is, as proposed by Parida et al. (2014), to invest in service innovation. Due to the cultural changes, Barquet et al. (2013) conclude that it is preferable to launch the service under a new business unit. Similarly, Mont et al. (2006) argue that starting a daughter company to manage the leasing service would reduce the risk of harming an already strong brand name that is closely connected to product innovation.

These understandings serve as a point of departure for an empirical study of the Scandinavian apparel presented in Sects 3.2 and 3.3.

3.2 Scandinavian Apparel: Empirical Background

The outdoor apparel industry is defined in this study as companies offering clothes, shoes, or any type of outdoor equipment. Products and services produced by the industry are supposed to equip consumers for outdoor activities, whether it is walking, skiing, or ice-skating; spending time in nature is key. Sustainability is therefore often a matter of course in the industry. Pelle (Pers. com. 2019) clarifies that Fjällräven's mission is to inspire people to spend more time in nature, which in itself is a sustainable practice since it has indirect positive effects on the environment. Studies indicate that once people experience outdoor adventures, they feel more connected to nature, which leads to more environmentally friendly lifestyle choices overall (Brymer & Gray, 2009; Boyes, 2013).

Gossen and Kropfeld (2022) suggest a transition in practices in the outdoor industry, which means that companies should adapt to influence society and the natural environment in a more controlled manner. The industry association recommends implementing monitoring practices to ensure ethically sourced products by considering animal welfare, labor treatment, and materials used in the value chain. Despite outdoor products generally being durable and of high quality (Hodgson, 2017), there are numerous challenges associated with materials used in production based on fossil fuels and chemicals (Bocken & Short, 2016). Hence, sustainability strategies within the outdoor apparel industry often consist of finding more eco-friendly materials that can deliver the same high level of performance. In this context, the concept of circularity is frequently discussed as a means to innovate products that are easily reused, repaired, and remanufactured (Mont et al., 2006; Azarenko et al., 2009; Hodgson, 2017).

The apparel industry is traditionally run by a take-make-dispose logic and is known as one of the world's most unsustainable industries with high pollution rates, water consumption, and chemical usage (Boström & Micheletti, 2016). Fast fashion trends motivate consumers to buy and throw away. The United Nations stated that the apparel industry generates more greenhouse gas emissions than all international flights and shipping together (Dory, 2018).

Clothes and high-quality equipment, such as many outdoor products, may not be a part of the fast fashion trend to the same extent, as these products are usually durable and designed to last a long time (Hodgson, 2017). However, Denise (Pers. com. 2019), the sustainability manager at Fjällräven, acknowledges that in a product-oriented business model, the customer still owns the decision rights over the product's use. Many outdoor products are season-based and are not used on a daily basis by most people. Durability and low utility rate are two product qualities that hold great potential for user-oriented services such as rental and leasing arrangements (Mont, 2002).

3.3 *Scandinavian Apparel: The Case Study*

A presentation of the collective case study is presented below, focusing on the kind of services offered, characteristics of a typical consumer, resources needed, financial aspects, material flows, and cultural changes. Each of these factors is synthesized in a use-oriented adapted version of the Canvas model.

3.3.1 Use-Oriented Services and Distribution Channels

The investigated companies have different use-oriented services, as illustrated in Table 3. Aktivt Uteliv opened their shop in Uppsala in 2014, offering rental and sales of a wide range of hard outdoor products, courses in various outdoor activities, and repairs in their workshop (Pers. com. Edvin 2019). Their rental concept, Rent-Try-Buy, intends to take customers all the way from an interest that emerges to rent a product, try it out, improve their techniques in a course, which may then lead to buying their own equipment because it has become their new hobby.

Respondents from established manufacturers Fjällräven, Bergans, and Houdini claim that they sell outdoor products to a wide range of internal and external retailers, but they have also recently started renting out products in their own brand stores. Anna (Pers. com. 2019) explains that in Bergans' flagship store in Oslo, customers are allowed to rent any product from the original assortment. Bergans also developed a special collection for outdoor activities in the Bessegen area, which is rented out through Gjendesheim's tourist cottage. Houdini has a similar offer of

Table 3 Use-oriented services and distribution channels in the studied outdoor retailer companies

Company	Type of company	Use-oriented service	Distribution channels
Aktivt Uteliv	Retailer Start-up (2014)	Short-term, monthly and seasonal rental	Store and Website
Bergans	Established manufacturer (1908)	Short-term rental	Brand store Oslo and Gjendesheim's cottage
Fjällräven	Established manufacturer (1960)	Short-term rental	Brand stores and Naturkompaniet
Houdini	Established manufacturer (1993)	Short-term rental	Brand stores
		Leasing subscription	Headquarters
Naturkompaniet	Retailer	Short-term	Majority of stores
Re:leased	Retailer Start-up (2019)	Leasing subscription and short-term rental	Website
Rent a Plagg	Retailer Start-up (2014)	Short-term, monthly and seasonal rental	Website and Travel Agents

short-term rentals of outdoor clothes in collaboration with Cooperhill Lodge in Åre (a ski-resort) where the rented clothes are delivered directly to customers' hotel rooms (Pers. com. Patrik 2019). Both companies agree that collaborating with a third party enables a wider distribution of the rental service. However, renting requires more work than selling, which can be challenging for a third party to handle, and the margins are bigger when using own channels. The companies would therefore like to lease out through their own website in the future, which would also generate a spin-off on other products such as wool base layers that are not as suitable for renting. Similarly, Naturkompaniet starts renting out machinery on a small scale in a selection of brand stores and through their internal retailer channel Naturkompaniet, as they plan to digitalize it in the future (Pers. com. Hans 2019).

The start-ups Re:leased, Rent a Plagg and Aktivt Uteliv offer rentals through their websites today. Jonna (Pers. com. 2019) acknowledges that although online shopping is becoming increasingly popular, retail cannot totally reject the physical meeting. She points out that: "The whole experience in a store needs to change drastically. Customers demand service and the possibility to try the clothes. They have no need to shop there." Rent a Plagg offers rentals through their website that can be picked up at the warehouse, in one of their delivery points around Åre, or delivered directly to customers' home (Pers. com. Markus 2019). The company collaborates with several travel agents that offer their rental service in combination with their trips.

3.3.2 A Typical Customer

Common features of the typical rental customer and their preferences are illustrated in Table 4. All respondents agree that rental offers attract customers who seldom use a certain product. For instance, customers who rarely use formal dresses suitable for special occasions such as birthdays, marriages, and graduations, or customers attending seasonal activities such as ski trips. Fjällrävens BM does not focus on

Table 4 Customer type and preferences

Customer type	Preferences
Seldom users and people participating in activities in different environments	Customization
Beginners with less willingness to pay	Lower prizes, access to premium quality, spreading knowledge, decreased entry barriers
Experienced and recurrent customers	Variation, trying new models and styles
Stressed families	Simplicity and solutions to ease their life situation
Younger generation	Ownerless consumption, less responsibilities for maintenance and sustainability

segmenting the market according to different customer groups (Pers. com. Pelle 2019). Instead, they distinguish between different environments where the products are used. For instance, their collection named Keb has technical features and ventilation possibilities suited for trekking at Kebnekaise. Similarly, Markus (Pers. com. 2019) argues that the biggest value proposition with rental is to offer customers a product that is specifically developed for their activity.

Rentals offer lower prices compared to buying, which enables customers to access premium products that they would not else afford to buy. Buying premium outdoor products requires a certain level of skill about how they should be used. Spreading knowledge, therefore, becomes a key value, and Hans (Pers. com. 2019) argues: "I see this as an opportunity to sell service in the form of selling knowledge. We could teach customers how to behave in nature, how to set up their tent, how to pack their backpack and how to light a liquor kitchen." Thus, the threshold for beginners to try outdoor activities is considerably lower through renting since consumers do not have to buy expensive products before they are sure that they actually enjoy it. Edvin (Pers. com. 2019) adds that rental does not only appeal to beginners but to experienced high-end consumers who are eager to try different models. Rental enables the use of a variety of products, and in the textile industry, this implies an opportunity for a varied and exciting wardrobe (Pers. com. Jonna 2019). It opens up for creativity and design to take place since customers are more likely to rent a piece of clothing that is a bit crazy.

In a pilot project to rent out snowsuits for children, Anna (Pers. com. 2019) discovered a great interest from stressed parents who did not have to spend time investigating which snowsuit is best in test, going shopping, and selling used clothes. Thus, rental services can be less time-consuming and solve practical everyday problems. This is an important aspect since customers have high demands for availability and convenience, and when owning products, it is the freedom to use them whenever you like that you buy. However, there is a cultural shift in new generations when it comes to the perception of ownership. Jonna (Pers. com. 2019) argues that younger people no longer have the need to own things. Instead, they desire "on-demand" and the ability to have access to all things at all times. Re:leased names this customer segment "The millennial," who want to support and identify themselves with companies that are transparent and have a greater purpose than only to generate profit. Ownerless consumption also implies that consumers don't have to worry about storage, washing, and repairing.

The interviewed companies describe their existing customer relationships as relation-based instead of transaction-based. Pelle (Pers. com. 2019) points out: "The relationship is thus not just about selling goods, but sharing experiences in nature." The idea with rentals is to build long-term dialogues with customers and receive valuable feedback. Through renting, companies withhold information about how much a specific product has been rented out, in what environment, how many times it has been washed, which customers rented what, etc. Thus, renting out is an opportunity to collect information about customer preferences to a greater extent than traditional retailers normally have access to.

3.3.3 Resources, Activities, and Partnerships

Like all linear businesses, one has to spend time and resources to get circular business ideas up and running. All respondents point out that a large financial capital is needed in a rental model since revenues are generated during a longer time period. Bergans investigates if they could develop depreciation rules or collaborate with a financial actor that could help bridge the income gap (Pers. com. Anna 2019). The current financial system is based on selling products, and credit institutions still use models that account for inventories that are quickly converted into money, not generated over time. Products become a rental business' key asset. Washing service and repairing require investing in multiple machines if it is conducted in-house. Moreover, space is needed in order to dry returned goods and check that they are in a good condition. Hans (Pers. com. 2019) believes that "It can be difficult to solve in every individual store because the retail space is very expensive and often limited in size." Furthermore, the after-treatment of products is time-consuming and likely requires extra staff. Moreover, Jonna (Pers. com. 2019) clarifies that offering rentals through a digital channel requires technological investments since e-comm is normally linear. Whether resources and activities are outsourced to partners is illustrated in Fig. 3.

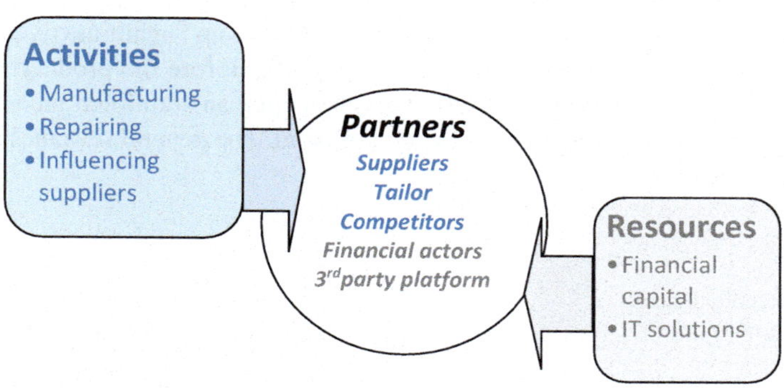

Fig. 3 Main factors influencing a use-oriented business model in terms of partners and resources, as interpreted from interviews

Fjällräven and Bergans are both manufacturers that produce high-quality and technically advanced outdoor products. Their core activity is, therefore, product development. However, a rental service requires a new set of activities, including after-treatment of products such as washing, drying, checking that they are in good condition, and repairing. When implementing circular BMs, Anna (Pers. com. 2019) argues that one has to find partners to a greater extent or even collaborate with competitors. If you do not have the knowledge and skills required in-house, you have to access it via your network and learn from others. Today, all companies in this study handle rentals and washing to a large extent in-house. Reparations, however, are commonly sent to a seamstress or back to suppliers, especially for products like hard-shell garments, since it requires advanced methods (Pers. com. Denise 2019). The reparations that can be done in-house are preferable since they increase the company's learning process and gain knowledge about how their products are worn out, which can later be used in product development.

3.3.4 Revenues and Costs

Revenue streams are generated in two separate ways. Firstly, some of the case companies offer leasing subscriptions that generate recurring income each month. Secondly, all companies offer short-term rentals or seasonal rentals that are paid at one point in time. Prices are either the same for all models or as a percentage of the original sales prices. For instance, Fjällräven charges 500 SEK to rent a tent during a weekend, while Bergans charges 10% of the price for the specific tent model. Generally, pricing is based on expected lifetime, purchase cost, and handling cost.

Common to all types of rental concepts is that the revenue streams are generated during a longer time period compared to selling products, which challenges companies to think long-term. Markus (Pers. com. 2019) argues that: "In the long run, this is a much more sustainable BM since we can earn more money on fewer goods. We are not at all as dependent on margins and seasons as a regular store." When the season is closed, traditional retailers arrange clearance sales to make room for new collections. On the contrary, rental BMs retain products for a longer period of time and simply "turn off" at the end of the season and "lit up" again next year. After a certain time, the product will generate 100% profit. Before the products are worn out completely, Patrik (Pers. com. 2019) explains that an additional income can be generated from selling the products as second-hand. The general revenues and costs are presented in Table 5.

Table 5 Reported factors influencing revenues and costs

Revenues +	Costs –
Recurring each month	Maintenance & storage
At one point in time	Freight
Sales of second hand	IT
Long-term revenues	Labor

Compared to selling, rental services imply added costs for storage, maintenance, extra staff, IT costs to integrate rentals into the website, and high freight costs since products will be returned. Elsa (Pers. com. 2019) points out that launching a new service is similar to launching a totally new BM. She perceives that the requirements are higher for services compared to products. Products are sometimes launched even if they are not profitable because they contribute to increased value for the brand.

3.3.5 Material Flows in the Value Chain

Many suppliers are skeptical towards renting since the apparel industry has struggled to adjust to different societal shifts such as customers' expectations on sustainable production, e-com and delivery-within-the-hour (Pers. com. Jonna 2019). Many actors do not see the economic potential with the model, and Markus (Pers. com. 2019) explains that: "We intrude on their way of doing business. Business is done in one way and it is about selling as many goods as possible and as soon as possible." From a retailer's perspective, it can be difficult to directly influence improvements in the product quality needed for rentals, especially if you are a small player. Rent a Plagg tries to change this pattern by involving their suppliers in the rental process so they can earn money during the rental period and not only at the point of purchase. This will give them incentives to produce products that last longer. Many of their suppliers have signed a cooperation agreement that states the suppliers take responsibility for maintaining the products.

In their current BM, Bergans strives to produce the highest possible quality at a reasonable price (Pers. com. Anna 2019). One has to take into account that customers must be willing to pay for quality. With rentals, there is an opportunity for manufacturers to produce products of extremely high quality. When Fjällräven started their pilot project, people worried that the products would be worn out quickly. Mikael (Pers. com. 2019) describes "..there is a challenge because the customer has less responsibility when she rents a thing compared to when she owns a thing. Therefore, I believe that the wear rate is higher when renting." Edvin (Pers. com. 2019) explains that it is a balance between striving for maximal quantities of rental cycles per product and ensuring that the product is of high performance. Aktivt Uteliv works actively to counter the wear issue through personal customer meetings, establishing a genuine family atmosphere, and trying to convey "take good care of our fine equipment." Their workshop repairs both rentals and private products, and the wear rate is roughly the same in both categories. However, among all the respondents in this study, there are contradictory views on whether wear and tear is a challenge with rentals.

In the near future, Bergans plans to launch a basic collection and a limited edition with increased quality specifically developed for renting (Pers. com. Anna 2019). Since they work closely with suppliers in product development today, their suppliers would probably be willing to increase the quality further. However, decreasing production volumes may be more challenging. Factories often have a requirement of a minimum order quantity, MOQ, since small orders entails too much work for the seamstresses. However, it would be easier for certain product types, such as tents, that are produced on a smaller scale and already struggle to reach the minimum quantities. One solution to reach the minimum quantities could be to order products less frequently and in low season when the fabrics are less busy. Decreasing sales volumes is, however, not always the objective for retailers that introduce rental services (Pers. com. Hans 2019). Naturkompaniet's main purpose is not to stop selling tents. On the contrary, they might be able to sell more tents since people can try different models before investing in their own.

3.3.6 Managing Reverse Logistics

There are logistical challenges with adopting a circular business model. Denise (Pers. com. 2019) points out that one must figure out where in the value chain it makes the most sense to provide rental services. On one hand, it would be favorable for the manufacturers if they offer rental products exclusively from their own brand. On the other hand, it might be more profitable for a retailer to provide the service since they could rent out many different brands and on a larger scale. It is then important that the retailer builds their own rental model that can integrate all brands. Patrik (Pers. com. 2019) argues that it is more efficient to keep products in few places, as the same products are used over and over again. Therefore, centralizing the rental services to a separate department could be beneficial. Both Bergans and Naturkompaniet's goal is to manage rental services from their central warehouses in Germany. However, Robin (Pers. com. 2019) points out that it is crucial that deliveries to and from a central warehouse are sustainable so that the entire climate gain with renting is not wasted. Therefore, Houdini considers transportation partners like Move By Bike for their leasing initiative "Subscription." In addition, they believe that the customer must get used to paying for the freight in order to put a value on the environmental costs of transportation. Since one must account for 100% returns, freight is expensive. Thus, finding good transportation partners is crucial.

In the linear model, the inventory is reduced the more customers that purchase from that company (Pers. com. Jonna 2019). In a rental model, it is the opposite. The physical inventory grows the more customers the company receives. Having a warehouse that grows with the customer base entails great logistical challenges. It also implies technological challenges since today's system is designed for the product to be purchased and then sold, not several transactions that a rental model requires. For all the established PO firms in this study, the linear IT system is perceived as a significant challenge to introduce rental services. Anna (Pers. com. 2019) states: "The challenge with renting out via our website is that the system is not built for the goods to come back." An alternative to invest in a new IT system is to cooperate with an external start-up that can build their technical platform based on rentals from the start. Moreover, new workplaces must be created in order for this circular BM to work. Several new services are needed such as service providers, technological solutions, logistics, and laundry. Jonna (Pers. com. 2019) points out that outsourcing washing and reparation requires a fast service since it creates an additional logistics chain. All the time the product is not rented out represents a missed income.

3.3.7 Cultural Change

Ever since the postwar period in Norway, Anna (Pers. com. 2019) describes that the norm has been to own things: "We are born and raised with the importance of owning." Rental implies a cultural change in the way we consume and how we perceive ownership. Moreover, it enables a business to disconnect profitability from production and consumption. Companies must, however, challenge consumers to accept shared ownership. Thus, there is a need to educate and influence customers to let go of ownership. Denise (Pers. com. 2019) experiences that rentals challenge companies in communicating a premium value since customers normally associate it with something less exclusive than owning. Products suitable for renting are expensive, require a lot of space for storage, and are used infrequently but are crucial to get a good experience. For instance, a bad tent and sleeping bag can ruin the whole experience of being in nature. Hygiene is another important factor, where some consumers believe that clothes that are worn close to the skin should be private and should not be shared with others. It is necessary for large players to take the step forward and offer rentals to be able to influence customer behavior to a greater extent. Edvin (Pers. com. 2019) expresses: "It takes an IKEA for people to understand that they can buy cheap furniture."

Cultural changes may also be needed internally. For a retailer like Naturkompaniet, Hans (Pers. com. 2019) does not believe that providing rental services would affect the organizational culture. Since employees are generally interested in sustainability, their attitudes would probably be positive unless the workload becomes too much. At Bergans, the majority of employees are proud to work for a company that takes responsibility for the environment and works actively with sustainable solutions. However, some people think that the company has to sell products in order to preserve the value of the brand. Anna (Pers. com. 2019) acknowledges that there are challenges to convince everyone about the economic potential of rental services and says that: "It an internal cultural change that we must do to bring everyone along." There is a need for increased knowledge in sustainable BMs and long-term profitability in order to convince people that renting is the way forward. Investing in service innovation is a way to signal the importance of services in the company, and Bergans will likely increase those resources in the future.

3.3.8 A Use-Oriented Perspective of the Canvas Model

Insights from the case study provide grounds for reinterpreting the Canvas model (Osterwalder & Pigneur, 2010, p. 40) with a use-oriented business model in mind (Fig. 4). The modified model provides an answer to the first research question about

<table>
<tr>
<td rowspan="2">Key Partners

Seamstress
Washing service
Financial actor
Service-provider
Competitors
Authorities</td>
<td>Key Resources

Human resources
Machines and space
IT system
Financial capital</td>
<td rowspan="2">Value Proposition

Low prices, high quality and access to knowledge (lower access barrier)

Customization, variation, sustainability and less responsibility for product lifecycle</td>
<td>Customer relationships

Long-term relationships, co-creation and co-production</td>
<td rowspan="2">Customer Segements

According to level of usage</td>
</tr>
<tr>
<td>Key Activities

Dry, wash and repair returned goods.
Influence customers to ownerless consumption.</td>
<td>Distribution Channels

Internal stores,
Provider at tourist destinations
3rd party digital provider,
Own website</td>
</tr>
<tr>
<td colspan="3">Cost Structures
Maintainance costs, staff and freight.
Investments in IT and products.</td>
<td colspan="2">Revenue streams
Transaction-based (short-term and seasonal),
Reccuring each month (subscriptions),
Long-term revenues and sales of second-hand</td>
</tr>
</table>

Fig. 4 Use-oriented business model perspective of the Canvas Business Model based on the case study of Swedish outdoor apparel

the characteristics of a use-oriented business model. The model is still static, which is a problem for understanding sustainability transitions, but some of the factors suggest altering the time perspectives in order to increase understandings of responsible value creation.

A reinterpreted use-oriented Canvas model (Fig. 4) suggests a more active role for the consumer in a relationship marketing perspective, a wider stakeholder engagement seen in more key partners, and newly identified key activities. It also shows different strategies concerning resource management, including financial resources. It is a reflection of a business strategy that is aimed at downstream CSR. These findings are further discussed below, focusing on three dimensions of a use-oriented PSS business model.

4 Discussion

Downstream CSR is characterized with a system perspective, a circular economy, where the consumer contributes to efficient resource use. CE advocates that products should spin in multiple loops of reuse, repair, and remanufacture (EMF, 2013). Thus, circular BM differs from linear models since materials are reused to a larger extent (Bocken et al., 2014). Besides the original elements in the canvas model, this study indicates that the UO BM includes three additional dimensions: reduced material flow, reverse logistics, and cultural adoption factors illustrated in Fig. 5.

PSSs are supposed to lead to reduced material flow in the value chain (Tukker, 2015). Since products are transported back and forth between companies and customers, it is important to manage reverse logistics (Mont et al., 2006). Moreover, changes in consumption patterns may require cultural changes (Barquet et al., 2013). These elements are explained in detail below (Sects. 4.1, 4.2, and 4.3).

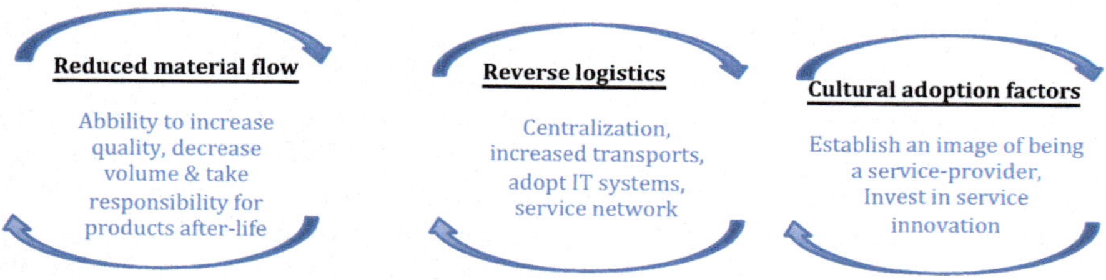

Fig. 5 Three additional dimensions to a use-oriented product system service business model

4.1 Reduced Material Flow

In contrast to linear business models, where firms are encouraged to maximize the number of units sold, user-oriented models create incentives to prolong the product life cycle by producing more durable goods that are easily maintained and remanufactured (Mont et al., 2006; Azarenko et al., 2009; Hodgson, 2017). However, the empirical results indicate that value chains in the outdoor apparel industry are to a large extent still managed according to linear logic and focused on selling as many goods as possible. Suppliers' attitudes to the rental concept are not always positive, as it intrudes on their way of doing business. To create incentives for manufacturers to increase the quality of their products, Rent-a-Plagg tries to involve their suppliers in the rental process through a cooperation agreement stating that the suppliers are responsible for maintaining the products. However, the general impression from manufacturing companies participating in this study is that they are positive towards improving product quality specifically suited for a rental service.

Since the responsibility of products shifts to producers in a user-oriented business model (Tukker, 2004), some of the companies in this study worry about wear and tear. The theoretical understanding is that UO BMs encourage companies to use the products as intensively as possible during its whole life cycle (Tukker, 2015). All respondents agree that there is a balance between striving for maximal utility of products and ensuring that the product is of high performance. Before products are worn out completely, they are likely to be sold as second hand.

Both the empirical and theoretical findings agree that a user-oriented business model breaks the link between profit and production volume. However, Bocken et al. (2014) highlight that the environmental benefits depend on the organization's ability to reduce production volume. Respondents from Fjällräven and Bergans claim that reducing production volume might be challenging. When ordering products from the factories, one has to consider minimum quantity requirements. Drastically reducing quantities on products that are sold large-scale is therefore not possible. Manufacturers may have to change their purchasing routines and order products less frequently, in low season and from fewer suppliers in order to reduce the production costs. From a retailer's perspective, however, decreasing sales volumes is not always the objective to introduce rental services (Pers. com. Hans 2019). Hence, this study indicates that the UO BM does neither automatically lead to increased product quality nor decreased production volume.

4.2 Reverse Logistics

When companies gain customers in a linear business model, the physical inventory decreases with each purchase (Pers. com. Norrman 2019). When renting, the warehouse increases as the customer base grows. This entails logistical challenges, both in physically managing the goods and technologically. Today, IT systems are designed for the product to be purchased and then sold. It is not designed to handle several transactions that a UO BM requires. This forces established companies to either invest heavily in IT to create a new platform and digitalize the process in-house or, as advocated by Tukker (2004), collaborate with a third-party service provider. However, respondents argue that rental services can be challenging for a third party to handle since they require a lot of work. Mont et al. (2006) clarify that it is important for the provider to manage reverse logistics. Hence, Norrman (Pers. com. 2019) argues that it is beneficial for manufacturers to collaborate with start-up retailers that build all their systems according to the reversed logistics from the start. Anna (Pers. com. 2019) perceives that Bergans' collaboration with the start-up service provider works well, and they consider continuing their collaboration to offer leasing subscriptions in the future.

As Denise (Pers. com. 2019) points out, there are few examples of product manufacturers that have transitioned to providing only services. In this shift, it is crucial to create a service network that links companies to customers and provides a route backwards from the user to the provider (Parida et al., 2014). An underdeveloped infrastructure in both of these two directions forces companies to extend their network and create new partners to compensate for the lack of in-house competencies and resources (Williams, 2007). When scaling up the UO services, the empirical results indicate that centralization is key. However, transportation to and from a central warehouse is a challenge since one has to account for 100% returns. This may result in high financial and environmental costs. Thus, Robin (Pers. com. 2019) points out that it is crucial to find sustainable transportation partners, for instance, by bike, and let customers pay for the freight. Anna (Pers. com. 2019) points out that new workplaces must be created since several new services are needed such as transportation, service providers, and technological solutions. However, despite the lack of infrastructure adapted to the reversed logics, companies must dare to try to get circular business ideas up and running. Anna (Pers. com. 2019) explains: "We do not just want to talk and analyze circular economy without testing it for real."

4.3 Cultural Adoption Factors

The norm, to own things, may imply barriers for a transition to rental related to, for example, trust regarding quality issues, maintenance, and hygiene of used goods (Rexfelt & Ornäs, 2009; Catulli, 2012); especially, soft products that are worn close to the skin and products that are associated with status and control (Tukker &

Tischner, 2006). To overcome these cultural barriers, it is crucial to establish the company as a service provider and in the eyes of the customer (Kindström, 2010; Reim et al., 2015; Armstrong et al., 2015), but communicating the values with a premium user-oriented service may therefore be challenging since customers normally associate rentals to something less exclusive than owning. The empirical result suggests collaborating with third-party providers and other partners to influence consumer behavior. Rent-a-Plagg, for instance, collaborates with travel agents who work closely with customers, and Markus (Pers. com. 2019) believes that they have greater abilities to change attitudes towards rental. Edvin (Pers. com. 2019) argues that it is necessary for large players to step forward in order to influence customer behavior to a greater extent. In agreement, Ceschin (2013) advocates that finding innovative solutions for a PSS may require collaborating with stakeholders who are not directly involved in the company. Since there is no global network for the development of sustainable services, respondents from Fjällräven and Bergans argue that companies may have to collaborate with competitors in order to find common solutions. Moreover, governments can provide legitimacy and support during the upscaling phase of eco-efficient innovations (Ceschin, 2013), and authorities have a significant role to influence consumer behavior.

For an established product-oriented firm to introduce services may induce resistance from actors that are involved in sales (Barquet et al., 2013; Armstrong et al., 2015). From a retailer's perspective, Hans (Pers. com. 2019) believes that unless the workload is too much, employees at Naturkompaniet are positive towards offering rental services since they are generally interested in sustainability. However, Anna (Pers. com. 2019) points out that some people believe that one has to sell products in order to preserve the value of the brand. She explains that investing in service innovation could increase the level of knowledge about rental services and convince employees about its financial potential. These findings support Mylan's (2014) understanding of PSS as a transition tool for sustainable development.

5 Conclusion

This chapter aims to identify enabling factors for downstream CSR in terms of a user-oriented business model. A user-oriented business model is characterized by long-term revenues that, in combination with high initial capital costs of purchasing products, result in requirements for financial resources. Moreover, customers are segmented according to the level of usage. A lower access barrier is a value proposition that attracts seldom users, while variation and customization attract frequent users. Since unique information about usage behavior can be collected through a user-oriented service, customers are perceived as co-inventors. To assure convenience and decrease transaction costs associated with booking, pick-up, and return, digital channels are beneficial. Delivering a user-oriented service implies a new set of core activities, including influencing customers in ownerless consumption, where partners and external service-providers are key. This study shows that except for changes in the nine elements of the Business Model Canvas, the user-oriented business model is characterized by three additional elements: the ability to reduce material flows, manage reverse logistics, and cultural adoption factors.

Secondly, an established product-oriented (PO) company within the outdoor apparel industry that introduces a use-oriented (UO) service may face several challenges. There is a long payback time in a UO BM, which can be difficult to manage financially. Due to doubts about profitability with a UO BM, internal resistance may rise, and companies may face challenges in managing cultural change. Companies may find it challenging to influence consumer behavior since the norm is still to own things. There are economic and environmental challenges to assure efficient transportation since products are delivered back and forth between companies and customers. Reverse logistics implies technological challenges since IT systems and e-commerce are normally linear. Moreover, there is a risk of rebound effects where products are worn out faster than privately owned products.

Finally, the result indicates that these challenges are met to a large extent through partnerships. The financial income gap due to long-term revenues may be supported by financial actors or by signing contracts with suppliers that allocate responsibilities, investments, and ownership rights. Multiple actors in society are needed to encourage ownerless consumption and to build the infrastructure needed to support reverse logistics. By delivering personal meetings to prevent wear and tear and choosing sustainable transportation partners, positive environmental gains can be assured. Moreover, companies in this study show that conducting pilot projects is key to get the circular business ideas up and running and to increase the organizational learning.

In conclusion, downstream CSR is a multifaceted approach that encourages businesses to consider their impact on society and the environment beyond just financial performance. It plays a vital role in fostering responsible and sustainable business practices, which serves as alternatives for consumers in today's interconnected and socially conscious world.

References

Ahn, Y., Okamoto, D., & Uejio, C. (2022). Investigating city bike rental usage and wet-bulb globe temperature. *International Journal of Biometeorology, 66*, 679–690. https://doi.org/10.1007/s00484-021-02227-5

Åkerlund, C. (2018). Hyrkläder ger hållbart mode. *Dagens industri*. Available from: https://www.di.se/hallbart-naringsliv/hyrklader-ger-hallbart-mode/. Accessed 20 Sept 2023.

Armstrong, C. M., Niinimäki, K., Kujala, K., Karell, E., & Lang, C. (2015). Sustainable product-service systems for clothing: Exploring consumer perceptions of consumption alternatives in Finland. *Journal of Cleaner Production, 97*, 30–39.

Azarenko, A., Roy, R., Shehab, E., & Tiwari, A. (2009). Technical product-service systems: Some implications for the machine tool industry. *Journal of Manufacturing Technology Management, 20*(5), 700–722.

Baines, T. S., Lightfoot, H. W., Evans, S., Neely, A., Greenough, R., Peppard, J., Roy, R., Shehab, E., Braganza, A., Tiwari, A., Alcock, J. R., Angus, J. P., Bastl, M., Cousens, A., Irving, P.,

Johnson, M., Kingston, J., Lockett, H., Martinez, V., Michele, P., Tranfield, D., Walton, I. M., & Wilson, H. (2007). State-of-the-art in product-service systems. *Proceedings of the Institution of Mechanical Engineers, Part B: Journal of Engineering Manufacture, 221*(10), 1543–1552.

Barquet, A. P. B., de Oliveira, M. G., Amigo, C. R., Cunha, V. P., & Rozenfeld, H. (2013). Employing the business model concept to support the adoption of product–service systems (PSS). *Industrial Marketing Management, 42*(5), 693–704. (Business Models – Exploring value drivers and the role of marketing).

Batat, W. (2019). Customer experience to keep up with changing consumer and new luxury consumption trends. In D. von Schweinitz (Ed.), *Kinderchirurgie* (pp. 55–74). Springer.

Bessant, J., & Davies, A. (2007). Managing service innovation. In *Innovation in services* (DTI Occasional Paper No. 9). Department of Trade and Industry.

Blüher, T., Riedelsheimer, T., Gogineni, S., & Klemichen., A & Stark, R. (2020). Systematic literature review – Effects of PSS on sustainability based on use case assessments. *Sustainability, 12*(17), 6989.

Bocken, N., & Short, S. (2016). Towards a sufficiency-driven business model: Experiences and opportunities. *Environmental Innovation and Societal Transitions, 18*, 41–61.

Bocken, N. M. P., Short, S. W., Rana, P., & Evans, S. (2014). A literature and practice review to develop sustainable business model archetypes. *Journal of Cleaner Production, 65*, 42–56.

Boehm, M., & Thomas, O. (2013). Looking beyond the rim of one's teacup: A multidisciplinary literature review of product-service systems in information systems, business management, and engineering & design. *Journal of Cleaner Production, 51*, 245–260.

Boström, M., & Micheletti, M. (2016). Introducing the sustainability challenge of textiles and clothing. *Journal of Consumer Policy, 39*(4), 367–375.

Boyes, M. (2013). Outdoor adventure and successful ageing. *Ageing and Society, 33*(4), 644–665.

Brymer, E., & Gray, T. (2009). Dancing with nature: Rhythm and harmony in extreme sport participation. *Journal of Adventure Education & Outdoor Learning, 9*(2), 135–149.

Carlsson, A., Hemström, K., Edborg, P., Stenmark, Å., & Sörme, L. (2011). *Kartläggning av mängder och flöden av textilavfall*. Sveriges Meteorologiska och Hydrologiska Institut. (SMED; 46 2011).

Catulli, M. (2012). What is uncertainty? Further insight into why consumers might be distrustful of product service systems. *Journal of Manufacturing Technology Management, 2*(6), 780–793.

Ceschin, F. (2013). Critical factors for implementing and diffusing sustainable product-service systems: Insights from innovation studies and companies' experiences. *Journal of Cleaner Production, 45*, 74–88. (Sustainable Innovation and Business Models).

Charter, M., & Polonsky, M. J. (1999). *Greener marketing: A global perspective on greening marketing practice*. Routledge.

Creswell, J. (2013). Five qualitative approaches to inquiry. In *Qualitative inquiry and research design: Choosing among five approaches*. SAGE.

Dory, K. (2018, June 27). Why fast fashion needs to slow down. *UN Environment*. Available at: http://www.unenvironment.org/news-and-stories/blog-post/why-fast-fashion-needs-slow-down. Accessed 20 Sept 2023.

Eidse, J. (2022). Rental's winners and losers. *Suston Magazine*. Available from: https://suston-magazine.com/2022/03/31/rentals-winner-and-losers/. Accessed 16 Aug 2022.

Eisenhardt, K. M. (1989). Building theories from case study research. *Academy of Management Review, 14*(4), 532–550.

El Alfy, A., Palaschuk, N., El-Bassiouny, D., Wilson, J., & Weber, O. (2020). Scoping the evolution of corporate social responsibility (CSR) research in the sustainable development goals (SDGs) era. *Sustainability, 12*(14), 5544.

Elo, S., Kääriäinen, M., Kanste, O., Pölkki, T., Utriainen, K., & Kyngäs, H. (2014). Qualitative content analysis: A focus on trustworthiness. *SAGE Open, 4*(1), 215824401452263.

EMF. (2013). *Circular economy report – Towards the circular economy* (Vol. 1). Ellen Macarthur Foundation. Available from: https://www.ellenmacarthurfoundation.org/publications/towards-the-circular-economy-vol-1-an-economic-and-business-rationale-for-an-accelerated-transition. Accessed 20 Sept 2023.

European Commission (EC). (2022). *EU strategy for sustainable and circular textiles.* https://ec.europa.eu/commission/presscorner/detail/en/QANDA_22_2015. Accessed 30 Oct 2022.

Gebauer, H., Fleisch, E., & Friedli, T. (2005). Overcoming the service paradox in manufacturing companies. *European Management Journal, 23*(1), 14–26.

Gossen, M., & Kropfedt, M. (2022). "Choose nature. Buy less". Exploring sufficiency-oriented marketing and consumption practices in the outdoor industry. *Sustainable Production and Consumption, 30*, 720–736.

Gunnarsson, M. (2019). *Use-oriented business models: A multiple case study of rental providers within the Scandinavian outdoor apparel industry*. Master Thesis report, Department of Economics, Swedish University of Agricultural Sciences. Available at: https://www.uppsatser.se/uppsats/de3da1a456/. Accessed 01 Jan 2023.

Hirschl, B., Konrad, W., & Scholl, G. (2003). New concepts in product use for sustainable consumption. *Journal of Cleaner Production, 11*(8), 873–881.

Hodgson, N. (2017). *Closing the loop on a circular economy*. Outdoor Industry Association. Available at: https://outdoorindustry.org/article/closing-loop-circular-economy/. Accessed 20 Sept 2013.

Karlsson, M., & Fernandes, C. (2016). Fast Fashion 2.0: en mer hållbar affärsmodell. Available at: http://urn.kb.se/resolve?urn=urn:nbn:se:hb:diva-10283. Accessed 20 Sept 2023.

Kindström, D. (2010). Towards a service-based business model – Key aspects for future competitive advantage. *European Management Journal, 28*(6), 479–490. (Management Focus: Managing Non-profit Organizations).

Klingelhöfer, C. (2018, March 13). *Ski and snowboard rentals are booming*. ISPO. Available from http://www.en/markets/rossignol-salomon-and-co-how-winter-sports-brands-view-rental-skis-and-snowboards. Accessed 12 Mar 2019.

Kónya, I., & Ohashi, H. (2004). *Globalization and consumption patterns among the OECD countries*. Social Science Research Network. (ID 558624).

Kuo, T. C. (2011). Simulation of purchase or rental decision-making based on product service system. *International Journal Advanced Manufactruring Technology, 52*, 1239–1249. https://doi.org/10.1007/s00170-010-2768-2

Kvale, S. (1996). *Interviews: An introduction to qualitative research interviewing*. SAGE.

Lewandowski, M. (2016). Designing the business models for circular economy – Towards the conceptual framework. *Sustainability, 8*(1), 43.

Lincoln, Y. S., & Guba, E. G. (1985). *Naturalistic inquiry*. SAGE.

Mark-Herbert, C., & Byfors, L. (2023). Consumer perspectives of rental of outdoor clothing. In S. S. Muthu (Ed.), *Novel sustainable alternative approaches for the textiles and fashion industry. Sustainable textiles: Production, processing, manufacturing & chemistry* (pp. 41–74). Springer.

Mark-Herbert, C., & Prejer, B. (2018). Social entrepreneurship in an inclusive business model. A new business model for sustainable agroforestry. In L. Mura (Ed.), *Entrepreneurship – Development tendencies and empirical approach* (Vol. 11, pp. 215–228). Intech. https://doi.org/10.5772/intechopen.70295. ISBN 978-953-51-3761-0, Print ISBN 978-953-51-3760-3.

McCoy, L., & Chi, T. (2022). Collaborative consumption: A study of sustainability presentation in fashion rental platforms. *Sustainability, 14*(14), 8537.

McEachern, M. G., Middleton, D., & Cassidy, T. (2020). Encouraging sustainable behaviour change via a social practice approach: A focus on apparel consumption practices. *Journal of Consumption Policy, 43*, 397–418. https://doi.org/10.1007/s10603-020-09454-0

Meier, H., Roy, R., & Seliger, G. (2010). Industrial product-service systems—IPS2. *CIRP Annals, 59*(2), 607–627.

Miles, M. B., & Huberman, A. M. (1994). *Qualitative data analysis: An expanded sourcebook*. SAGE.

Mont, O. (2002). Clarifying the concept of product–service system. *Journal of Cleaner Production, 10*(3), 237–245.

Mont, O. (2004). *Product-service systems: Panacea or myth?* Dissertation Lund University. Available from: http://lup.lub.lu.se/record/467248. Accessed 22 Sept 2023.

Mont, O., Dalhammar, C., & Jacobsson, N. (2006). A new business model for baby prams based on leasing and product remanufacturing. *Journal of Cleaner Production, 14*(17), 1509–1518. (Product Service Systems: reviewing achievements and refining the research agenda).

Moro, S. R., Cauchick-Miguel, P. A. & de Sousa Mendes, G. H. (2022). Adding sustainable value in product-service systems business models design: A conceptual review towards a framework proposal. *Sustainable Production and Consumption, 32*, 492–504.

Muthu, S. (2017). Evaluation of sustainability in textile industry. In S. Muthu (Ed.), *Sustainability in the textile industry. Textile science and clothing technology* (pp. 9–15). Springer. https://doi.org/10.1007/978-981-10-2639-3

Myin, M., Su, J. W., & H & Shen, H. (2022). Investigating the determinants of using clothing subscription rental servies: A perspective form Chinese young consumers. *Young Consumers, 24*(1), 95–113.

Mylan, J. (2014). Understanding the diffusion of sustainable product-service systems: Insights from the sociology and consumption and practice theory. *Journal of Cleaner Production, 97*, 13–20.

Novick, G. (2008). Is there a bias against telephone interviews in qualitative research? *Research in Nursing & Health, 31*(4), 391–398.

Oliva, R., & Kallenberg, R. (2003). Managing the transition from products to services. *International Journal of Service Industry Management, 14*(2), 160–172.

Osterwalder, A., & Pigneur, Y. (2010). *Business model generation: A handbook for visionaries, game changers, and challengers*. Hoboken, New Jersey, Wiley.

Parida, V., Sjödin, D. R., Wincent, J., & Kohtamäki, M. (2014). Mastering the transition to product-service provision: Insights into business models, learning activities, and capabilities. *Research-Technology Management, 57*(3), 44–52.

Reim, W., Parida, V., & Örtqvist, D. (2015). Product–Service Systems (PSS) business models and tactics – A systematic literature review. *Journal of Cleaner Production, 97*, 61–75. (Special Volume: Why have 'Sustainable Product-Service Systems' not been widely implemented?).

Rexfelt, O., & Ornäs, V. (2009). Consumer acceptance of product-service systems: Designing for relative advantages and uncertainty reductions. *Journal of Manufacturing Technology Management, 20*(5), 674–699.

Schoonover, H., Mont, O., & Lehner, M. (2021). Exploring barriers to implementing product-service systems for home furnishing. *Journal of Cleaner Production, 295*, 126286.

Schor, J. (2010). *Plenitude: The new economics of true wealth*. Penguin Press.

Shuk, S. W. (2016, April 21). *Fast fashion is "drowning" the world. We need a fashion revolution!* Greenpeace International. Available from: https://www.greenpeace.org/international/story/7539/fast-fashion-is-drowning-the-world-we-need-a-fashion-revolution. Accessed 20 Sept 2023.

Tojo, N. N. C. O. M. (2012). *Prevention of textile waste: Material flows of textiles in three nordic countries and … suggestions on policy instruments*. Nordic Council of Ministers.

Tu, L., & Ngo, M. (2018). *Ett styrverktyg för implementering av tjänstefokuserade affärsmodeller – Balanserat styrkort*. Available from: https://gupea.ub.gu.se/handle/2077/56810. Accessed 29 Apr 2019.

Tukker, A. (2004). Eight types of product–service system: Eight ways to sustainability? Experiences from SusProNet. *Business Strategy and the Environment, 13*(4), 246–260.

Tukker, A. (2015). Product services for a resource-efficient and circular economy – A review. *Journal of Cleaner Production, 97*, 76–91. (Special Volume: Why have 'Sustainable Product-Service Systems' not been widely implemented?).

Tukker, A., & Tischner, U. (2006). Product-services as a research field: Past, present and future. Reflections from a decade of research. *Journal of Cleaner Production, 14*(17), 1552–1556. (Product Service Systems: reviewing achievements and refining the research agenda).

Tunn, V. S. C., Van den Hende, E. A., Bocken, N. M. P., & Schoormans, J. P. L. (2021). Consumer adoption of acccess-based product-service systems: The influence of duration of use and type of product. *Business Strategy and the Environment, 30*(6), 2796–2813.

VR, Swedish Research Council. (2017). Good research practice. VR1710, ISBN 978-91-7307-354-7. https://www.vr.se/download/18.5639980c162791bbfe697882/1555334908942/Good-Research-Practice_VR_2017.pdf. Accessed 09 Sept 2023.

Williams, A. (2007). Product service systems in the automobile industry: Contribution to system innovation? *Journal of Cleaner Production, 15*(11–12), 1093–1103.

Witell, L., Snyder, H., Gustafsson, A., Formbelle, P., & Kristensson, P. (2016). Defining service innovation: A review and synthesis. *Journal of Business Research, 69*(8), 2863–2872. https://doi.org/10.1016/j.jbusres.2015.12.055

Yin, R. K. (2013). *Case study research: Design and methods* (5th ed.). SAGE.

CSR Activities of Apparel and Fashion Brands: An Indian Context

R. Surjit, M. Shubha Shree, E. Shalini, and B. Anusri Mahalakshmi

Abstract Corporate social responsibility is an approach or a business initiative taken to improve societal well-being through business practice and the contribution of corporate resources. The main aim of this initiative is to achieve the welfare of the community through their social development activities. The issues related to CSR have been the subject of growing debate across social sciences, business, and management studies. Countries like China have been facing pressure to take CSR issues seriously, especially regarding environmental and labor standards. India became the first country to introduce the Companies Act 2013, which has made it obligatory for certain specified companies to grant at least 2% of their net profits to social welfare efforts. Performing CSR activities helps organizations create an optimistic influence among the public, and it promotes a healthier relationship with customers. Many Indian companies are actively indulged in CSR, among which Reliance Industries, Tata Consultancies and Services, and HDFC Bank Limited have funded around 1000 crores for this initiative. In the case of Indian apparel and fashion brands, Shopper's Stop, Arvind Brands, Aditya Birla Retail, Raymond, and Lifestyle perform CSR activities and are committed to sustainable production practices as part of CSR. The focus areas of CSR from these organizations include education, rural transformation, human rights, marginalized groups improvement, health, sports, and development, etc. A portion of their funds for CSR activities is allocated to public welfare activities and also to government-promoted schemes. In the year 2019, the project Su. Re, i.e., Sustainable Resolution, was launched in association with the Clothing Manufacturers Association of India, IMG Reliance, and the United Nations in India. The top 16 Indian apparel brands are the signatories to the project. This project's main motive was to establish a sustainable pathway for the fashion industry of India. Thus, CSR establishes a new way of doing business that creates value through a respectful and proactive attitude among stakeholders.

R. Surjit (✉) · M. S. Shree · E. Shalini · B. A. Mahalakshmi
Department of Fashion Technology, PSG College of Technology,
Coimbatore, Tamil Nadu, India
e-mail: rst.fashion@psgtech.ac.in

S. S. Muthu (ed.), *Corporate Social Responsibility in Textiles and Fashion*,
Sustainable Textiles: Production, Processing, Manufacturing & Chemistry,
https://doi.org/10.1007/978-3-031-61099-8_3

In this chapter, the various CSR activities of Indian brands are to be detailed along with the scope for improvement.

Keywords Indian brands · CSR activities · Sustainability · Fashion clothing · CMA

1 Introduction

Corporate social responsibility (CSR) is all about the enterprises behaving as responsible entities in society by contributing a portion of their profit to the betterment of the community, country, and aiming for sustainable development. It aims to create an ultimate positive impact on the environment. CSR activities have the potential to position the brand positively as consumers are becoming more aware of this subject in the present. As CSR is considered one of the critical aspects in running a brand or a company, knowing its components and having a closer look at these is considered important for a better understanding of the concept. In this chapter, the pillars of CSR are discussed for a deeper understanding of this concept. CSR is a step that businesses take for the betterment of society and to create an impact beyond profits. Many companies hire consulting agencies to better prioritize their social activities. The apparel and fashion industries, being the second-largest pollutants on the earth, have a responsibility to contribute in turn to the society that acts beneficial to them. This chapter will focus on the overview of CSR concepts and their importance in the current period. India consists of a large number of apparel brands and clothing industries ranging from luxury to affordable brands. Many apparel companies, such as Aditya Birla, Arvind, etc., are contributing to social welfare through various projects, which are discussed in detail in this chapter. Sustainability is one of the important issues in CSR in the apparel industry. Lots of sustainable and slow fashion brands like Ka Sha, Doodlage, etc., are emerging in India intending to create products that inflict no harm to the environment. In this chapter, the details about CSR, the pillars of CSR, types of CSR including environmental, economic, philanthropic, and ethical CSR, the importance of CSR, CSR activities of apparel companies, challenges faced in adopting CSR activities in apparel industries, the role of the Indian government in CSR activities, the role of Indian brands in CSR activities, the impact of CSR on Indian companies, and the scope of CSR are discussed elaborately.

2 CSR: Corporate Social Responsibility

Corporate social responsibility is defined by the World Business Council for Sustainable Development as “CSR is the continuing commitment by business to behave ethically and contribute to economic development while improving the quality of life of the workforce and their families as well as of the local community and society at large." (Moir, 2001). CSR is a model that comprises the activities or initiatives taken by a company to be accountable to both stakeholders and society. It aims to operate in a manner that involves creating a positive impact on the environment and providing contributions that enrich society. The main concept behind CSR is to bring a changeover among the companies from mere compliance to the mode of engagement. CSR is all about the sense-making of organizations, and the character of the organization is understood in depth through three process dimensions (Cognitive, Linguistic, and Conative). The Cognitive dimension involves the identity of the organization understood through the shared perceptions and legitimacy to be achieved by the organizations. The linguistic dimension involves the justification of the actions carried out by the organizations, which infer both the relationship with stakeholders and societal responsibilities, as well as transparency in communication, which can be balanced or biased. Conative dimension involves responsive postures, consistency between the organization's strategy and its activities related to CSR, and the organization's commitment (Basu & Palazzo, 2008). CSR communication is done by companies through their annual reports, which help position their brand among consumers (Lindgreen & Swaen, 2010).

3 Pillars of CSR

The CSR theories are based on the core idea of the triple bottom line concept, which is considered to be the three primary principles of corporate social responsibilities. According to the concept of the triple bottom line, the company has to take responsibility for the three attributes: Profit (economy), People (society), and Planet (environment), also known as the three P's of CSR (Księżak, 2016; Księżak & Fischbach, 2017). As all three attributes are closely related to each other, they have to be taken care of equally and should have a balanced effort. If any one attribute is not looked at properly, it may lead to an issue. For example, if the company looks only for profit and the welfare of the people and has abandoned and omitted the environment's well-being, it would destroy the place where we live (earth). Considering only the people and planet by the companies could be good to hear, but a company obviously would require profits for the better run of the company in the future and for sustaining. On the other hand, if it focuses only on the planet and profit but not on people, then it will lead to a reduction in the contraction of the societal reach.

3.1 Profit

This constitutes the economic aspect of the CSR. It focuses not only on the increase and improvements in profits of the company but also plans for the better use of the profits earned and their contribution to the social economy. The economic responsibility of the company relies on three aspects namely:

- The multiplier effect: It works on the aspect that the economic performance of the company is found to have both direct and indirect effects on all its employees, customers, local organizations, etc.
- Contribution through taxes.
- Eliminating activities that damage the trust (Uddin et al., 2008).

3.2 People

This attribute of CSR starts from the facilitation and support provided to the workforce of the company. This is done by encouraging the employees with frequent awards, medical leaves, exemptions for valid reasons, and improvement in workplace ergonomics. It will not only make them more loyal to the company but also raise productivity. The responsibilities of the companies towards people are categorized by Uddin et al. (2008) into three aspects: Responsibility towards customers, employees, and the community. The other aspect considered important is equitable treatment of the employees and customers irrespective of their age, gender, caste, community, etc. Creating a better perception of the company among the local communities is said to be another crucial aspect that helps in gaining its position against its competitors.

3.3 Planet

The natural environment and its betterment are considered to be the responsibility of every person on Earth, and primarily initiatives are taken by the corporate companies in this regard. The corporates are sometimes found to be the primary means of polluting the environment and causing impact and damage to society by irresponsible extensive usage of available resources. So, to counter the damages caused during the production of the product, the company has to eliminate or reduce the detrimental effects caused by it to the environment. Corporations have implemented sustainable methods of producing products, assessing the lifecycle of the product, etc. Some sustainable practices followed by clothing companies include the use of organically processed cotton, regenerated biodegradable fibers produced from regenerated cellulosic fibers like modal, lyocell, etc., and the use of polylactic acid fiber whose raw material is renewable, nonpolluting, compostable and more environmentally friendly. Many companies have moved on to using sustainable and renewable fibers for production. For example, H&M which is a fast fashion brand states that it uses certified organic cotton, implements many recycling programs, and replaces polyurethane with environment-friendly water-based substitutes as a part of sustainable activity thereby strengthening its CSR cause.

4 Types of CSR

Corporate Social Responsibilities are classified as follows:

- Environmental responsibility
- Economic responsibility
- Philanthropic responsibility
- Ethical responsibility

4.1 Environmental Responsibility

In recent years, many industries and organizations have been paying attention to the impact created on the environment and are focusing on implementing some practices to reduce the impact created on the environment. Many citizens, environmental organizations, and leadership companies define corporate environmental responsibility as the duty to cover the environmental implications of the company's operations, products, and facilities; eliminate waste and emissions; maximize the efficiency and productivity of its resources; and minimize practices that might adversely affect the enjoyment of the country's resources by future generations (Mazurkiewicz, 2004).

4.2 Economic Responsibility

Economic responsibility is an important type of CSR. It is followed by manufacturing firms to spend or invest their finances in doing good for society or the environment. The revived economic Corporate Social Responsibility has a moral independent of investor property rights. Economic responsibility argues that no costly responsibility action should be undertaken voluntarily (Windsor, 2006). To understand the economic phenomenon of CSR, questions like these should be asked:

1. Whether economic responsibility exists?
2. When and to what extent it can be efficient?
3. When it should exist? (Kitzmueller & Shimshack, 2012)

The goal of economic responsibility is to create a good environment for the people and society and not focus on increasing their profits.

4.3 Philanthropic Responsibility

The philanthropic responsibilities of a firm refer to the voluntary donations and services done by an industry or firm to the environment and society. This can be done by either donating or doing voluntary work for the benefit of society. These activities also create a good brand name for the firm. The main aim of the philanthropic responsibility is to create a good impact in society. Some firms create their organizations and trusts to improve the environment and society. Thus, philanthropy is the voluntary part of a company or organization, even if there are always some expectations from society in this regard (Grigore, 2010).

4.4 *Ethical Responsibility*

The ethical responsibility of an organization is the responsibility of the business to operate its firm on fair terms by standing by its principles. This includes fairness in the sourcing of materials, providing enough compensation and wages for employees, as well as treating them with respect. In an ethical organization, each individual is considered to be equal. It can also be considered as those responsibilities including the new values and norms that society expects to be fulfilled by the enterprise, even if such values and norms can reflect higher performance standards than those provided by law (Grigore, 2010).

5 Importance of CSR

The projects associated with CSR are beneficial to both society and the company as they serve to fulfill societal needs and also help to build a positive brand image. Corporate Social Responsibility is found to be one of the most representative drivers of Customer-Company Identification (CCI). As per the literature, CSR has a great influence on loyalty behavior among customers, which is likely more favorable to the company (Huang et al., 2017). Based on a study conducted by Grover (2014), CSR has a positive impact on the brand image of the company. Publicizing the CSR activities of the company is essential as it acts as an effective marketing approach for market share expansion. For example, the erstwhile Aircel company's share rose 10% after their high promotion of the "Save Our Tigers" campaign (Grover, 2014). Another important context of CSR is the commitment of the clothing industries in developing countries toward better working conditions of the laborers and the way they deal with sweatshops (Thorisdottir & Johannsdottir, 2020). There lies a strong relationship between CSR and consumers, as consumers are becoming more aware of this concept. It's been observed that there is a willingness among consumers to pay more for an ethical product, and if the company is found to be socially irresponsible, then the consumption rate of the products would eventually decrease (Gupta & Hodges, 2012). The right strategy for CSR can be formulated by the companies' such that the economic and social goals converge (Lindgreen & Swaen, 2010).

6 CSR in Apparel Companies

One of the crucial contributors to global issues is apparel companies. As clothing is again one of the basic everyday requirements, the apparel industry was among the first industries that were under pressure to adopt the activities of CSR at all points of the supply chain.

6.1 Challenges of Aadopting CSR in Apparel Industries

Fashion and textile products are major, crucial elements of daily life and are considered basic elements in creating identity and social relationships. On the other hand, apparel companies are the industries that have expanded production all over the world, and with cheaper labor, they can make considerable profits for the economy of the nation. It creates hesitation for the companies to alter and regulate the supply chain (White et al., 2017). The next challenge would be to involve everyone in the company or orient everyone towards the common cause of social and environmental welfare activities. The employees and the management would be reluctant to carry out CSR activities due to the involvement of more money and time. Another challenge is to align the company vision, mission, and values concerning CSR as non-alignment may lead to diffusion or dilution of CSR activities as they are not part of the company's values. Many companies may see this as a potential opportunity for increasing profits and may superficially do CSR activities. It should be done more genuinely with care and concern to ensure success and development. Some companies have limited availability of resources at their disposal, and using them for CSR activities will affect their regular business. All these cause the reluctance of the companies to perform CSR activities to the best possible extent.

6.2 Apparel Companies

6.2.1 Aditya Birla Retailers

From the CSR policy reports of Aditya Birla, we infer that they are an active, socially responsible company that majorly focuses on girl child education, adult education programs, rural developments, health incentives for stakeholders, sustainable livelihood (by forming self-help groups, women empowerment groups, agricultural development through better focus on farmers, partnership with training institutes), infrastructure development, etc.

Every manufacturing unit of Aditya Birla has a CSR cell, and they have allocated a predetermined amount from the profits for the CSR activities. In collaboration with ABG (Aditya Birla group) and FICCI, they have set up CSR centers to excel and make CSR an integral unit of corporate culture. It has also stepped into sustainable practices by measuring its water, waste, energy, and greenhouse gas emissions. In 2015 and 2016, they achieved energy reduction by incorporating some measures which included LED installation and improved operational efficiency. Speaking about the sourcing and production aspects, they stick to the usage of nonhazardous chemicals like eco resins, production of eco-friendly products such as zero detergent shirts, and the usage of reusable plastic crates along with a reduction of 130 t of packaging. Lastly, on the technological aspect, IT software like ENABLON is used by them for managing their sustainable activities.

6.2.2 Raymond

Raymond's four major focus areas are rural area development, eradicating hunger, poverty, and malnutrition, healthcare, and environmental sustainability. They have always focused on developing the lower sections of society, and for this, Raymond Embryo Research Centre researches embryo transfer in cattle. They have specifically come up with a mission that helps in improving the rural area through cattle breed development called JK Trust Gram Vikas Yojana. This mission has contributed to the reduction of infant mortality and malnutrition among children by the surplus milk production. This initiative gave hands and benefitted over 45 million rural people across 45,000 villages in 11 states of India.

Raymond has a rehabilitation center that focuses on around 16-year-old underprivileged teenagers, providing vocational classes conducted by faculties from industrial training institutes, which help these children to become self-reliant in every aspect. Concerning their community initiatives, Raymond has taken its unique initiative by starting its skilled tailoring institute to train all the unemployed women, less privileged people, and minority communities in the art and science of tailoring, which in turn enhances the skill set, increases the employment, and also promotes young entrepreneurship.

6.2.3 Fabindia

They are more focused on linking rural producers and urban markets. They focus on implementing vocational skills for rural people. They have a school named Fabindia in Rajasthan managed by Bhadrajun Artisans Trust, in which 45% of the students are girls. This school aims to provide high-quality education, especially for rural children. The company is closely engaged with the Artisans Cluster Development Trust (ACDT), a registered trust for the welfare and betterment of the artisans and craftsmen community. ACDT was formed to improve the lives of people in lower sections of society, with more specific concentration on aspects like education, financial assistance, sanitation, healthcare, etc. Fabindia has recorded spending about 2% of its net profits for CSR activities, as per their CSR reports.

7 Indian Government Roles in CSR

The Indian Government has an important role in Corporate Social Responsibility; it ensures that all organizations follow the rules and norms set by society. To develop CSR practices, the government can create partnerships with businesses to direct them in a way that leads to the betterment of society. For the improvement of the private sector, the government is said to play a major role in reducing the cost, reducing the risks associated with the process, and opening the door for private firms and sectors for their betterment. The constraints faced by the government in promoting CSR are to identify the areas that need to be prioritized, increase awareness, develop incentives and support, and mobilize resources from cross-sectorial cooperation, which are a support for building on existing initiatives and capacities. Some key roles that a government can actively choose to engage in to support a CSR agenda are listed below:

- Through regulation, it can come in the form of laws, regulations, penalties, and associated measures to control the aspects of business investment or operations.
- Government can inspect the business behavior by setting some targets to achieve, developing a few standards to improve the performance of a firm, and implementing environmentally friendly systems. Establishing a minimum age for labor forces can be an example of this.
- Government can direct the firms or industries to engage in CSR activities for the improvement of the environment and society. This can be done by educating industries on CSR activities.
- To promote ethical business in society, the government can provide incentives on tax and other duties to the firms indulging in CSR activities and make them understand CSR and its impact on society.
- To overcome environmental and social difficulties, the government can play the role of a mediator between businesses, people, and other organizations. As a mediator, the government can also involve itself in promoting Corporate Social Responsibility. Educating awareness among society, and offering funds for research, campaigns, etc. can be examples of this.
- The government can help companies perform CSR activities in various ways like educating companies on CSR activities done internationally and also explaining the principles of international businesses on CSR activities. They can also conduct various programs with the prime aim of creating awareness and publicizing CSR activities.
- The government can also provide awards for good CSR activities.

Some initiatives that can be taken by the government to improve CSR activities in the country are as follows:

- Around the world, many governments have started implementing Corporate Social Responsibility activities. In an enabling environment, new policies can be introduced in society, and their effectiveness can be measured.

- In a democratic country, CSR cannot be made mandatory against the interests of the firm. So, to promote CSR, the benefits of it should be taught to all firms and enterprises.
- Awareness about how CSR plays an important role in the country's growth should be educated to the people by the government.
- The first step to promoting CSR in a country is to create awareness and knowledge about the significance and contribution of CSR to business success and sustainability, as well as to increase acceptance (Stiftung & GTZ, 2007).
- Recently, many governments have started setting up organizations to promote CSR activities in society. For example, in 2007, a CSR institute was established in Thailand under SET to increase awareness about CSR among the people.
- To implement good CSR practices and performances, the government plays an important role. If there is any constraint for a business practice by law or tax, the government has the authority to change it.
- Governments can reform regulatory frameworks to meet standards related to Corporate Social Responsibility.
- Governments interact with NGOs, enterprises, and other organizations to implement CSR activities and also reduce the problems created through these activities.
- In many ways, the government creates partnerships with enterprises and organizations to initiate and develop CSR activities.

8 Indian Brands and Their Roles in CSR

The fashion industry is one of the booming sectors in India. India is growing to be the major apparel producer among all other countries in the world. There are brands of wide varieties situated in India ranging from fast fashion like Zudio to slow fashion like Aavaran, luxury brands like Sabyasachi to affordable fashion brands. Various fashion brands, both large and small scale, are present that focus on many categories in apparel ranging from ethnic wear to sportswear, footwear, and accessories, providing a wide range of unique collections for men, women, and kids.

CSR is deeply connected with Indian culture and tradition. During the period of industrial changes, Indian organizations have realized the importance of public contribution to fulfill societal needs. Many policies were implemented by the companies to address the issues related to labor, community, and the environment (Lattemann et al., 2009). It is also observed that the main areas of focus for CSR activities by the apparel industries of India are education, healthcare, and welfare schemes for employees and their families (Gupta & Hodges, 2012). The fashion and apparel industries, being the second-largest contributor to environmental pollution, are trying many sustainable approaches and CSR practices seriously to safeguard the environment (Hency, 2019). The various activities carried out by them are discussed below:

8.1 *Activities by Fashion Brands*

The main aspect of CSR activities provided by the industries is to dedicate a certain portion of their profit to the schemes beneficial to both society and the environment. In specific, the apparel industries are doing a lot of activities that contribute to the welfare of the community. The main focus areas of CSR activities are:

- Healthcare
- Education
- Employee welfare
- Women empowerment
- Environmental sustainability
- Rural and slum area development
- Disaster management
- Financial support to needy students, and much more

8.1.1 Healthcare

Many fashion brands are known to contribute a part of their CSR fund to the healthcare sector as their CSR activity, conducting various programs that ensure the welfare of people's health. Especially during the pandemic, the government of India requested all the companies to keep their predominant CSR focus activities on healthcare and disaster management. About 5 crore INR contribution was made by the Reliance group for the relief fund of the Maharashtra Chief Minister (Tanya, 2020). Another notable group that has made a significant contribution to the healthcare community is the Aditya Birla group. They run 24 medical hospitals and have conducted 6000 medical camps with the main objective of providing good facilities to villages and elsewhere. They are also known to conduct mother and child care projects, immunization programs, awareness programs, and healthcare facilities for physically challenged people. Arvind Lifestyle Brand Ltd. has planned to establish medical camps in association with the Swasth India Foundation as a part of their CSR projects with a mission of providing quality healthcare facilities to the community at a lower cost. Several other brands like Raymond, Fabindia, etc. are known for their significant contribution to this sector.

8.1.2 Education

Education is one of the most vital needs of the community. It becomes essential for the growth and welfare of the society. Hence, almost all companies try to spend a portion of their CSR fund on this sector. For example, Arvind Lifestyle Brands Ltd., one of the leading textile manufacturers, is known for improving the infrastructure of schools by partnering with the local governments. They have created a flagship program called Gyanda and the main objective of this concept is to provide basic supplementary education for children studying in the Municipal schools. The Aditya Birla Scholarships, instituted in 1999 by the Aditya Birla group, provide significant financial support to students who are willing to pursue higher studies. For the welfare of underprivileged children, vocational courses are provided by the Raymond Rehabilitation Center to make these students more financially independent.

8.1.3 Women Empowerment

Women empowerment is another important segment for the welfare of society. Nowadays, many apparel companies are aiming to provide necessary programs that are beneficial to women's empowerment. One of the main commitments of the Anita Dongre Foundation is towards women's empowerment. They have created an initiative of setting up Community Tailoring Units in the rural areas of Maharashtra to provide professional training in garment making to the needy women in those areas. Fabindia, an Indian chain store, is also known to participate in activities that promote gender equality and facilitate the employment of women to make them socially empowered.

8.1.4 Sustainability

In the present, the term "Sustainability" has taken a wider reach, and many sustainable approaches are adopted by all the current fashion industries for the betterment of the environment. Sustainability's main objective is to meet the current needs without compromising the requirements of the future generation to meet their own needs. Exploring the connection between CSR and sustainability, it's been observed that the term sustainability is described to come under the umbrella of Corporate Social Responsibility (Thorisdottir & Johannsdottir, 2020). Hence, the fashion industries need to focus their CSR activities on the three different aspects of sustainability, which are environmental sustainability, social sustainability, and economic sustainability (Ozek, 2017), as one of the high polluters of the world is the textile and apparel industry. The sustainability practices can be adopted by industries through various techniques like Upcycling, Recycling, Reconstruction, and Zero-waste approach. An Indian brand, No Nasties, is known to produce organic and vegan clothing, and it ensures that the products produced by it offset the carbon footprints. Another brand, Sui, is a slow fashion brand whose products are made of natural fibers and colored using herbal dyes. Doodlage is a brand that upcycles factory waste into amazing limited-edition products and their packaging is also ensured to be plastic-free (Jharna, 2023; Doodlage, 2023). Amit Aggarwal has also created a line of products under his AM.IT design label using recycled plastics and other uncommon materials. Likewise, many other brands like 11.11 (Eleven Eleven), Nicobar, Ka-Sha, etc. are well-known for their sustainable designs.

8.1.5 Rural and Slum Area Development

About 65% of the Indian population residing in rural areas is not developed. Hence, many companies are focusing their contribution on rural upgradation and the welfare of the community. The Anita Dongre Foundation is known to provide skill training and create livelihood opportunities for women in rural areas. Okhai, an Indian fashion brand, creates its products with intricate detailing through underprivileged women artisans from rural regions, thereby creating employment opportunities for them. Arvind Lifestyle Brand Ltd. aims to provide clean, hygienic drinking water to the people living in slum areas by working together with the Ahmedabad Municipal Corporation. They have also created safer houses in partnership with the Ahmedabad Municipal Corporation for the slum dwellers of Sanjay Nagar located in Ahmedabad to stay in.

8.1.6 Disaster Management

One of the main foci among the CSR projects would be disaster management. Companies lend their helping hand by contributing a portion of their profit to the relief funds, thereby helping the community overcome difficult situations. For example, Shopper's Stop, an Indian department store chain, spent 10 lakhs for the Chennai flood relief fund. Many big fashion brands have made significant contributions to disaster relief as part of CSR activity.

9 Impact of CSR on Indian Companies

CSR has proven beneficial in increasing the company's profile at local, national, and international levels. The performance of CSR by companies has not only impacted the particular year but also the following years. The CSR performance by the companies has been different from one company to the other (Garg, 2016). Through The Companies Act 2013, it has come into existence that companies should spend 2% of their profit on CSR activities. CSR spending has been very low compared to the percentage of revenue and profits during the voluntary spending period (Verma & Kumar, 2014).

In India, companies like Tata Group have been involved in CSR activities like developing rural areas, providing education to children, and healthcare services through the Tata Index for Sustainable Development (Tata Sustainability Group) (Tata Group, 2012). Many other companies like Wipro, Infosys, etc. have also adopted CSR activities. A study by Arora and Puranik (2004) found that CSR activities in India were mostly concentrated around the sectors of community development and more or less not around the sustainability focus of CSR activities. The philosophy of CSR has not been properly defined in many companies or enterprises. A study by Arevalo and Aravind (2011) concluded that Indian companies are driven by moral motives rather than profit motives. A study by Verma and Neha (2012), without any references, states that the CSR spending of companies in India for the year 2009–10 was 7.5 billion. A study by Verma and Kumar (2014) concludes that compulsory employing of 2% profit of a company on CSR activities is acceptable to a certain extent looking at the CSR expenditure of well-established companies in the past.

10 Scope of CSR

CSR is said to be one of the key factors in determining the customer and company relationship. Socially responsible activities done by companies tend to attract different ranges of customers towards the company. Thus, the implementation of an activity or program in the right way determines the type and amount of customers the company attracts. Service quality, in turn, denotes the company's ability to produce and service well. If the company excels in providing excellent service, customers will be ready to exploit new opportunities and build a stronger bond with the brand. The relationship between a customer and a company is dynamic (Huang et al., 2017).

Implementation of CSR not only helps in customer and company relationships but also in many other aspects like creating a valued culture in the organization. As many of the socially responsible activities are focused on health, education, staff, and customer welfare, it is said to help in the betterment of society indirectly and create a corporate wokeness. Wokeness is a sense of awareness that the organizations have toward social problems and inequality. CSR is said to be commonplace because of two main reasons. Firstly, there are increasing expectations among the stakeholders and the customers about the company's privacy and policies aligned according to the principles of sustainability. Next, CSR is said to contribute more to the tourism and hospitality organizations, and this, in turn, improves customer satisfaction, increases labor productivity, and financial performance, thus helping in embracing the corporate wokeness.

11 Conclusion

Corporate Social Responsibility benefits both society and the brand, as through CSR projects, many social and environmental issues can be addressed, and through effective marketing of their activities, the growth of the company also rises. People are more aware in the present related to this topic, hence CSR has a high influence on the loyalty of the consumers. Indian fashion brands are spending their CSR fund on healthcare, education, women empowerment, slum rehabilitation, etc., which has benefitted lots of people in need. Due to the growing development of CSR, it is important to know the idea of a socially responsible company, one such approach is the triple bottom line, though it was developed years ago, it is considered to be the ideal method to apply as it mainly insists in the coexistence of profit, people, and planet. Corporate Social Responsibility has proven to be more beneficial to a company's profile at local, national, and international levels. Indian government tends to play an important role in improving CSR activities across the country. In upcoming years, the awareness about CSR among consumers will increase, and the success rate will also tend to increase, thereby leading to the benefit of society and the country.

References

Arevalo, J. A., & Aravind, D. (2011). Corporate social responsibility practices in India: Approach, drivers, and barriers. *Corporate Governance, 11*(4), 399–414.
Arora, B., & Puranik, R. (2004). A review of corporate social responsibility in India. *Development, 47*(3), 93–100.
Basu, K., & Palazzo, G. (2008). Corporate social responsibility: A process model of sensemaking. *Academy of Management Review, 33*(1), 122–136.
Doodlage. (2023). https://doodlage.in/ [online]. Accessed 23 Oct 2023.
Garg, P. (2016). CSR and corporate performance: Evidence from India. *Decision, 43*, 333–349.
Grigore, G. (2010). Ethical and philanthropic responsibilities in practice. *University of Petroşani~ Economics, 10*(3), 167–174.
Grover, A. (2014). Importance of CSR in inclusive development. *Procedia-Social and Behavioral Sciences, 157*, 103–108.
Gupta, M., & Hodges, N. (2012). Corporate social responsibility in the apparel industry. *Journal of Fashion Marketing and Management: An International Journal, 16*(2), 216–233.
Hency, T. (2019). 7 Fashion brands that take CSR seriously. *The CSR Journal.* [Online]. https://thecsrjournal.in/seven-fashion-brands-that-takes-csr-seriously/. Accessed 24 Oct 2023.
Huang, M. H., Cheng, Z. H., & Chen, I. C. (2017). The importance of CSR in forming customer–company identification and long-term loyalty. *Journal of Services Marketing, 31*(1), 63–72.
Jharna, P. (2023). [Online]. https://consciousfashion.co/guides/india-sustainable-fashion-brands. Accessed 23 Oct 2023.
Kitzmueller, M., & Shimshack, J. (2012). Economic perspectives on corporate social responsibility. *Journal of Economic Literature, 50*(1), 51–84.
KsiężaK, P. (2016). The benefits from CSR for a company and society. *Journal of Corporate Responsibility and Leadership, 3*(4), 53–65.
KsiężaK, P., & FischBach, B. (2017). Triple bottom line: The pillars of CSR. *Journal of Corporate Responsibility and Leadership, 4*(3), 95–110.
Lattemann, C., Fetscherin, M., Alon, I., Li, S., & Schneider, A. M. (2009). CSR communication intensity in Chinese and Indian multinational companies. *Corporate Governance: An International Review, 17*(4), 426–442.
Lindgreen, A., & Swaen, V. (2010). Corporate social responsibility. *International Journal of Management Reviews, 12*(1), 1–7.
Mazurkiewicz, P. (2004). Corporate environmental responsibility: Is a common CSR framework possible. *World Bank, 2*(1), 1–18.
Moir, L. (2001). What do we mean by corporate social responsibility? *Corporate Governance: The International Journal of Business in Society, 1*(2), 16–22.
Ozek, H. Z. (2017). Sustainability: Increasing impact on textile and apparel industry. *Journal of Textile Engineering & Fashion Technology, 2*(5), 506–509.
Stiftung, B., & GTZ. (2007). *The CSR navigator: Public policies in Africa, the Americas, Asia and Europe. Partner for the future worldwide.* Bertelsmann Stiftung.

Tanya, K. (2020). https://in.apparelresources.com/business-news/sustainability/retail-brands-stay-committed-society-csr-campaigns-covid-19/. [Online]. Accessed 22 Oct 2023.

Tata Group. (2012). Towards improving the quality of life of the people. Retrieved from https://www.tatasustainability.com/SocialAndHumanCapital/CSR#:~:text=The%20Tata%20group's%20activities%20relate,very%20purpose%20of%20its%20existence. [Online]. Accessed 21 Oct 2023.

Thorisdottir, T. S., & Johannsdottir, L. (2020). Corporate social responsibility influencing sustainability within the fashion industry. A systematic review. *Sustainability, 12*(21), 9167.

Uddin, M. B., Tarique, K. M., & Hassan, M. R. (2008). Three dimensional aspects of corporate social responsibility. *Journal of Business and Economics, 3*(1), 200–212.

Verma, A., & Kumar, C. V. (2014). An analysis of CSR expenditure by Indian companies. *Indian Journal of Corporate Governance, 7*(2), 82–94.

Verma, R., & Neha, M. (2012). Corporate social responsibility: A study of DMRC. *Review of Research, 1*(11), 1–4.

White, C. L., Nielsen, A. E., & Valentini, C. (2017). CSR research in the apparel industry: A quantitative and qualitative review of existing literature. *Corporate Social Responsibility and Environmental Management, 24*(5), 382–394.

Windsor, D. (2006). Corporate social responsibility: Three key approaches. *Journal of Management Studies, 43*(1), 93–114.

Gender Influences at the Core of CSR Initiatives and Sustainable Fashion

G. D. Muller, N. C. Sonnenberg, and S. Donoghue

Abstract Women have always fulfilled a central role in fashion and have been the principal focus of those who create apparel. Sustainable Development Goal 5 (SDG 5), which seeks to promote gender equality and empower women, is therefore well-ingrained in the traditional makeup of the clothing and textile industry. Yet, this industry is evolving, and a gender-neutral approach is increasingly advocated. With gender-fluid fashion set to gain more traction in years to come, the future of gender-based marketing and, more specifically, gender differentiation in clothing retailers' Corporate Social Responsibility (CSR) campaigns is questioned. The debates become even more pronounced in emerging markets where traditional views of gender may still prevail. By definition, CSR entails business-specific initiatives and practices aligned with social, environmental, and economic dimensions based on the Triple Bottom Line (TBL) assumptions. Consumers' perceptions of CSR and their support for CSR initiatives are critical in pursuing a sustainable clothing and textile industry. Research has shown that gender is an essential determinant of environmental and socially responsible clothing consumption behaviors and, by implication, engagement in clothing retailers' CSR initiatives. This chapter explores topics surrounding SDG 5 and its intersection with the fashion industry, particularly within emerging markets, gender differentiation, and gender neutrality in apparel consumption, as well as differences in male and female consumers' perceptions of CSR. The chapter concludes with a theoretical framework and research propositions for future investigations. The social-cognitive theoretical assumptions that underlie gender influence and the implications for gender-based marketing, segmentation, and differentiation in CSR campaigns, particularly in emerging market contexts, are discussed.

G. D. Muller · N. C. Sonnenberg (✉) · S. Donoghue
Department of Consumer and Food Sciences, University of Pretoria, Pretoria, South Africa
e-mail: niel.muller@up.ac.za; nadine.sonnenberg@up.ac.za; sune.donoghue@up.ac.za

S. S. Muthu (ed.), *Corporate Social Responsibility in Textiles and Fashion*, Sustainable Textiles: Production, Processing, Manufacturing & Chemistry,
https://doi.org/10.1007/978-3-031-61099-8_4

Keywords Corporate social responsibility · Triple bottom line · Gender-based marketing · Gender-fluid fashion · Gender differentiation · Emerging markets · Sustainable clothing · Consumer perceptions

1 Introduction

The textile and clothing industry, a major pillar of the global economy, has been progressively undergoing a vital transformation toward sustainability within the realms of Corporate Social Responsibility (CSR). The industry's operational practices have been scrutinized due to their substantial environmental impact and numerous ethical issues, which has hastened the need to adopt sustainable practices that are in harmony with the Triple Bottom Line (TBL). TBL promotes the incorporation of social, environmental, and economic considerations into business plans. Beyond business plans, a deeper examination of the sector unveils the complexity of consumer behavior, which is shaped by diverse sociocultural factors, with gender dynamics being particularly influential.

Gender, as a crucial societal construct, has a striking influence on consumption patterns, contributing to the diversification of sustainable behaviors and perceptions across different consumer groups. The interplay between gender and sustainability within consumer behavior is subtle yet influenced by many factors, including societal norms and stereotypes. Distinct gender roles and expectations often shape individuals' attitudes toward sustainability and CSR initiatives, guiding their preferences and purchase intentions within the context of fashion and textile consumption.

With the above in mind, this chapter provides an overview of key topics surrounding gender influences, including SDG 5 and the role of women in molding the clothing and textile industry. The chapter is structured as follows: Sect. 2 explores traditional gender role perspectives. Building upon this foundation, Sect. 3 provides a comprehensive overview of Sustainable Development Goal 5 (SDG 5) and its role in promoting gender equality and empowering women. Section 4 then takes a historical perspective, exploring the instrumental role of women in shaping the clothing and textile manufacturing sector. This historical context sets the stage for Sect. 5, which examines CSR through the Triple Bottom Line (TBL) and its three dimensions: social, economic, and environmental. Section 6 navigates the intricacies of consumers' perceptions, explicitly focusing on discerning differences between male and female perspectives. The narrative progresses to Sect. 7, which investigates the relevance of gender marketing in the clothing sector and its potential influence on consumer choices. Next, Sect. 8 reviews the contemporary shift toward gender-neutral fashion. The comprehensive discussion culminates in Sect. 9, which offers a conceptual framework and future research recommendations. Section 10 concludes the chapter and advocates for applying gender theories, such as social identity theory and gender schema theory, to interpret and contextualize gender influence. At their core, these theories offer a basis to examine CSR initiatives aiming to advance sustainable fashion practices. These theories emphasize the need to view gender not just as a dichotomous variable but as a complex entity deeply linked to societal norms and individual identities. This understanding is vital for developing tailored CSR strategies that can efficiently drive sustainable fashion choices, potentially creating a deeper bond with the customer base through embracing or challenging established gender schemas.

2 Traditional Gender Role Perspectives

In defining gender, it is essential to differentiate between biological sex, which refers to physiological attributes, and gender, which encompasses socially constructed roles and identities (Fausto-Sterling, 2000; Kaiser, 1997; Ryan, 2007). In other words, while "sex" denotes the biological classification as male or female, "gender" encapsulates societal perceptions and roles. Butler (2011) further emphasizes the performative nature of gender, suggesting its construction through societal and individual actions.

Deeply rooted in historical, cultural, and socioeconomic contexts, traditional gender roles define men's and women's expectations, responsibilities, and behaviors (Kaiser, 1997). The origins of traditional gender roles can be traced back to religious doctrines, legal systems, and cultural traditions (Kaiser, 1997). Traditionally, women have been relegated to caregiving and performing household tasks, whereas men have predominantly been considered the family's primary earners (Eisenberg & Lennon, 1983; Wood & Eagly, 2012). Historically, the stringent binary perception of male and female roles has led to power imbalances, positioning men predominantly in authoritative roles in politics, the economy, and family dynamics. In contrast, female roles have been limited to the domestic sphere, reducing women's opportunities for formal education and professional growth. These restrictions curtailed their potential and led to socioeconomic inequalities that, in some cases, persist (Sen, 2001). Grau and Zotos (2016) comment on various studies that reveal ongoing widespread gender stereotyping, yet other studies, such as those of Eisend (2010) as well as Bhatia and Bhatia (2021), confirm a substantial decrease in the intensity of gender stereotyping in recent times.

Sustainable Development Goal 5 (SDG 5), with its emphasis on gender equality, has brought traditional gender roles and gender stereotyping into the spotlight, challenging societies to reconsider the foundational basis of gender-specific roles. Specific targets within this goal include the assurance of women's complete involvement in leadership positions and the abolition of harmful practices grounded in gender prejudices (United Nations, 2015). Over several years, authors like Sen (2001) have emphasized the need to eliminate harmful practices rooted in gender bias and establish an environment where men and women can thrive without predetermined role-based restrictions. Women increasingly occupy roles traditionally reserved for men, such as heads of state, CEOs of corporations, or primary breadwinners of families. Simultaneously, men challenge stereotypes by taking on caregiving roles, breaking away from toxic masculinity, and supporting shared responsibilities in domestic spheres (Bittman et al., 2003; Eisend, 2010; Grau & Zotos, 2016). With global movements and objectives such as SDG 5 influencing societies, a gradual shift toward gender equality occurs, and new paradigms emerge.

3 SDG 5: Promoting Gender Equality and Empowering Women

SDG 5 forms part of the 17 global goals set out by the United Nations in their 2030 Agenda for Sustainable Development across three key areas: economic, social, and environmental, as well as realizing gender equality and empowering all women and ensuring human rights for all. The overarching ambition of SDG 5 is to achieve gender equality and empower all women and girls. The persistent and historical challenges that women and girls encounter globally, including discrimination and violence, as well as limiting their political, economic, and social participation, emphasize the critical nature of this objective (United Nations, 2015).

SDG 5 defines several targets, which serve as measurable objectives to achieve its main ambition. For example, to end all forms of discrimination (Target 5.1) and violence (Target 5.2) against women and girls and to abolish harmful customs (Target 5.3). In 1993, the United Nations General Assembly adopted the Declaration on Elimination of Violence against Women, whereby the declaration defined gender-based violence as "any act that results in or is likely to result in physical, sexual, or psychological harm or suffering to women, including threats of such acts, coercion or arbitrary deprivation of liberty, whether occurring in public or private life" (United Nations, 2008). In addition to constituting severe human rights violations, gender violence carries discernible economic repercussions. As such, it restricts women's engagement in the labor market, reduces their productivity owing to physical and psychological distress, and imposes supplementary strains on healthcare infrastructures, translating to substantial economic costs that could amount to as much as 3.7% of a country's GDP (Thawre & Singh, 2021). The effects of gender-based violence extend beyond the immediate victims as it fosters a continuous cycle of discrimination and disparity, affecting future generations' societal cohesion. SDG 5 accentuates the critical role of reforms, consciousness-raising campaigns, and sturdy legal frameworks in stopping violence against women and girls in all spheres of life. Furthermore, it points out the imperative of setting up services to aid victims and systems to ensure the accountability of those who perpetrate such violence (United Nations, 2015).

SDG 5 Target 5.4 acknowledges and values unpaid care and domestic labor through delivering public services, infrastructure, and supportive policies. Target 5.5 advocates for women's inclusive and impartial participation in leadership and decision-making roles across different domains. Target 5.6 accentuates the crucial nature of facilitating universal access to sexual and reproductive health services and rights. Target 5.a calls for substantial reforms to ensure women have equal rights to economic resources and opportunities to control assets and properties. In addition, Target 5.b promotes enabling technology to advance women's empowerment. Lastly, Target 5.c concludes with a call for robust and enforceable policies and legislation to nurture gender equality, safeguard rights, and empower all women and girls at every societal tier (United Nations, 2015). Holistically, SDG 5 seeks to develop societies where gender does not dictate one's rights, opportunities, or privileges, steering toward a future of equal opportunities and justice for all.

With its global outreach and pervasive influence on culture and lifestyle, the fashion industry shapes gender perceptions, behaviors, identities, and norms worldwide. Given the breadth of its impact, the sector holds considerable power in either presenting or even challenging traditional gender norms (Crane, 2012). Gender-specific trends and roles historically dominated the fashion industry, which may have significantly influenced SDG 5's targets.

Historically, the fashion industry has both empowered and exploited women. On the one hand, it has offered employment opportunities for millions of women worldwide, especially in the garment production sector (Ruwanpura, 2011). On the other hand, it is also notorious for underpaying, overworking, and exploiting these same women through unfavorable working conditions and limited opportunities for upward mobility (Dickson & Eckman, 2006; Ruwanpura, 2011). Therefore, focusing on gender equality propels the industry to reconsider its practices, from ensuring fair wages to safe working environments for women.

4 The Role of Women in the History and Evolution of the Clothing and Textile Manufacturing Sector

Historically, and to date, the clothing and textile industry workforce is predominantly composed of women (Molly & Larner, 2010; Vijeyarasa & Liu, 2022). Women have been central to apparel creation, weaving, embroidery, and other intricate textile processes for centuries (Chadwick & Chadwick, 1990). In ancient societies, such as those of Greece and Rome, women were predominantly engaged in domestic duties, with textile production a crucial aspect of these activities. Specifically, weaving was considered a home-based activity, and numerous artifacts from those periods portray women as the primary individuals involved in weaving (Barber, 1994).

As the world entered the nineteenth century, the Industrial Revolution significantly changed the textile industry (De Vries, 1994), with many women employed in textile mills and garment factories (Lim, 1990; Ruwanpura, 2011). While mechanization made production faster, it also brought along challenges for women. Despite being the backbone of the industry, women have faced systemic wage disparities and, in many cases, challenging working conditions such as long hours and minimal wages (Lim, 1990). Despite technological advancements and shifts in societal attitudes, modern-day textile factories, especially in developing countries, still rely heavily on female labor. Women continue to face challenges such as wage disparities, lack of representation in leadership roles, and, in some cases, unsafe working environments (Ross, 2004). The collapse of the Rana Plaza building in 2013 in Dhaka, Bangladesh, which caused the death of 1132 garment workers (mainly women), serves as a reminder for stakeholders in the clothing and textile sector to address these inequalities in their business practices (Hobson, 2013).

Elkington (1997) emphasizes the characteristics a socially sustainable organization should uphold, including equitable labor practices, community engagement, fair allocation of human resources, and a steadfast commitment to social justice. In the realm of CSR, these issues form part of the social dimension of sustainable endeavors, with organizations supporting local communities, including female members of those communities, based on ethical and moral foundations (Öberseder et al., 2014; Schwartz & Carroll, 2003).

5 Corporate Social Responsibility (CSR)

CSR is a multifaceted concept with various definitions, causing scholars and practitioners to constantly pursue a universally endorsed definition (Carroll, 1991; Russo & Perrini, 2010). The foundational concept of CSR, set forth by Bowen (1953), suggests that business entities should tailor their policies, decisions, and pursuits to align with societal objectives and values. In the 1960s, Keith Davis proposed an expansive perspective on social responsibility, indicating that it transcends an organization's immediate economic or technical scope (Carroll, 1991). During this period, CSR was envisioned as a means to address the intricate dynamics arising from corporations' influence on society and lay down ethical principles governing the relationship between corporations and society (Eells & Walton, 1961, as cited in Carroll, 1991). This narrative evolved, advocating an integrated view where business and society are interlinked rather than distinct (Moir, 2001; Wood, 1991).

Additionally, an emerging perspective emphasizes that while the main objective of organizations is to fulfill societal needs by providing goods and services, fostering a relationship with society is crucial for maintaining a stable environment (Cannon, 1992; Moir, 2001). Kotler and Lee (2005) depicted CSR as an initiative to enhance community welfare through elective business practices and contributions of organizational resources. Hence, CSR can be interpreted as companies' manifestation of responsible behavior, going beyond legal requirements and addressing their impact on society, the environment, and the supply chain (Jansen van Rensburg, 2016).

Through the years, the concept of CSR has matured, embodying a voluntary combination of social and environmental aspects into business strategies (Lillywhite & Diviney, 2007). CSR now encompasses a broad range of issues, including community investment, environmental stewardship, philanthropy, workers' rights and welfare, anti-corruption measures, corporate governance, and market relations, to name a few (Blowfield & Murray, 2014; White et al., 2017). Beyond these aspects, governmental regulations also shape CSR, where organizations must adhere to legislation (Jansen van Rensburg, 2016), including policies, programs, structures, and mechanisms to empower women and transform gender relations in all aspects of work (Das, 2022).

Based on findings from Aguinis and Glavas (2012), CSR at the organizational level rests on five pivotal conclusions. Firstly, companies involve themselves in CSR mainly for instrumental reasons, focusing on anticipated financial gains. Secondly, firms' CSR engagement is also driven by normative motivations rooted in the company's values, representing a commitment to doing what is morally right. Thirdly, a slight yet positive relationship exists between CSR initiatives and financial achievements. Even though the actual scale of the CSR–financial outcomes relationship remains inconclusive, numerous nonfinancial advantages emerge, encompassing enhanced management practices, better product quality, increased operational efficiencies, appeal to investors, and boosted demographic diversity, including the representation of women and ethnic minorities. The fourth conclusion relates to the opportunities that surround managerial perceptions of CSR, which may underscore the intangible resources of a firm. Finally, the relationship between CSR and outcomes intensifies with high exposure and visibility and when the company's size is substantial.

Despite subtle nuances in organizational approaches, CSR may be viewed as a company's sustained commitment to enhancing economic, social, and environmental well-being (of both men and women) by utilizing its resources, policies, and business operations (Du et al., 2011). These principles closely align with the assumptions of the Triple Bottom Line.

5.1 Adopting the Triple Bottom Line (TBL)

On a global scale, there is a noticeable shift in the focus of retailers and corporations, moving toward addressing social and environmental concerns and expanding beyond purely economic goals such as profit-making (Tsarenko et al., 2013). From a business perspective, achieving sustainable development involves balancing social, environmental, and financial aspects, constituting the Triple Bottom Line (TBL) framework (Elkington, 1997; Henry et al., 2019). The TBL is a progressive business framework conceived by John Elkington in 1997, proposing that companies should emphasize social and environmental performance equally as they do financial performance, creating a trifold focus on "people, planet, and profit." Elkington (1997) argues that a balanced emphasis on environmental and social considerations alongside economic gains may foster sustainable development that satisfies current needs without jeopardizing future generations' ability to fulfill theirs. This approach prompts organizations to reflect on the comprehensive implications of their operations and scrutinize their influence on a diverse array of stakeholders such as employees, local communities, and the environment, moving beyond an exclusive focus on shareholders and fiscal results (Elkington, 1997). The TBL approach essentially measures organizational sustainability performance in the three domains, namely the social, environmental, and economic dimensions of CSR (Elkington, 1997).

5.2 The Social Dimension of CSR

Öberseder et al. (2014), in their comprehensive scale development and validation study, identify various dimensions of the CSR social/societal domain, specifically from a consumer perspective. These include the employment of disabled and long-term unemployed individuals, educational advancements, and addressing pressing societal issues. In a separate scale development study, additional societal dimensions were identified, such as endorsing public health and educational initiatives, adhering to well-defined ethical values, financially supporting social endeavors, and fostering an enhanced quality of life within communities (Alvarado-Herrera et al., 2017). Specifically in the textiles and clothing industry, societal considerations are intrinsically linked to the provision of fair wages to workers, including women, ensuring safe and healthy working environments, eradicating child labor, promoting local production, and contributing to community development (Jägel et al., 2012; Lichtenstein et al., 2004; Park & Kim, 2016). Studies conducted in emerging market contexts such as South Africa highlight a notable emphasis on the social dimension, potentially due to the implementation of government-mandated regulations such as Broad-Based Black Economic Empowerment (BBBEE) on local organizations (Dawkins & Ngunjiri, 2008; Mersham & Skinner, 2016). This emphasis is particularly poignant given the alarming unemployment rate of 44.1% in 2022 in South Africa, underscoring the significance of the social dimension in addressing one of the country's most critical challenges (Statistics South Africa, 2022).

5.3 The Economic Dimension of CSR

The economic domain emphasizes fiscal stability through competitiveness, employment generation, and sustained profitability (Jamali, 2006). Organizations can achieve these financial goals by curbing operational costs through sensible resource use and elevating productivity via a motivated workforce (Jamali, 2006). In their study, Alvarado-Herrera et al. (2017) pinpointed six distinct facets of the economic aspect, which entail profit maximization for enduring sustainability, cultivating powerful customer relationships, the constant upgrading of service quality, implementing competitive pricing strategies, enhancing fiscal outcomes, and consistently increasing productivity. From a consumer's perspective, the economic domain relates more specifically to the organization's financial health and product quality

(Fulton & Lee, 2013; Gruber et al., 2014). In selecting garments, consumers' economic considerations may, for example, touch upon attributes such as functionality, longevity, comfort, design, and quality of materials and craftsmanship in clothing (Jägel et al., 2012; Park & Kim, 2016).

Focusing on the broader implications of the economic dimension, it may be particularly relevant in elevating women to men's level of socioeconomic status in emerging economies such as South Africa. While South Africa transitioned to a democratic governance structure in the mid-90 s, its economic trajectory has stalled lately. Factors such as political volatility combined with the devastating effects of the COVID-19 pandemic led to an unprecedented spike in poverty rates and the highest inflation in over a decade (The World Bank Group, 2022). It is also worth mentioning that in February 2023, the Financial Action Task Force (FATF) greylisted South Africa due to its insufficient measures against money laundering and the financing of terrorism (Perumall, 2023). Therefore, the economic domain is essential in the country for the general public and specifically for women.

5.4 *The Environmental Dimension of CSR*

The environmental domain of CSR focuses on an enterprise's ethical resource utilization and its endeavors to lessen its environmental footprint (Alhaddi, 2015). Such endeavors could include energy conservation, effective waste and emission management, resource optimization, and endorsing eco-friendly commodities or procuring locally produced items (Goel (2010). In emphasizing a firm's environmentally responsible operations, Öberseder et al. (2014) and Alvarado-Herrera et al. (2017) underline activities such as curbing energy consumption and emissions, managing waste, advocating for recycling initiatives, bankrolling environmental conservation research and development, backing green initiatives, and consuming predominantly natural resources. The clothing industry mirrors the environmental domain's dimensions by promoting green manufacturing processes, sidestepping harmful chemicals, designing products that have minimal environmental impact and cause no harm to animals, and preferring sustainable fabrics such as organic cotton (Jägel et al., 2012; Park & Kim, 2016; Shen et al., 2012). In this regard, previous studies have substantiated women's consistent alignment toward environmentally friendly behaviors compared to men (Davidson & Freudenburg, 1996; Zelezny et al., 2000). Females often display more environmental concern than men (Dunlap, 1983) and exhibit higher ethical standards (Simionescu, 2015). Furthermore, women are more active in practicing eco-conscious activities across developed and developing nations (Vicente-Molina et al., 2013).

From the discussions above, it can be deduced that companies embracing the TBL framework are committed to social responsibility, environmental care, and robust economic results. They integrate social and environmental considerations into their decision-making processes, exceeding the basic legal requirements (if any are required). By doing so, they focus on bettering community ties, elevating employee well-being, including that of female employees, and lessening environmental harm (Aguinis & Glavas, 2012). According to Elkington (1997), businesses benefit from an improved public image, heightened employee contentment, and diminished regulatory scrutiny through adopting TBL principles. Additionally, this approach motivates enterprises to innovate their offerings, addressing societal and ecological issues, thus unlocking new market avenues. The growing prominence of CSR has a ripple effect on consumer preferences, with an increasing number of consumers incorporating social and environmental considerations into their purchasing decisions Pérez (2009). This has led to heightened awareness and a preference for products and services offered by socially and environmentally responsible organizations or those adhering to CSR principles (Hilson, 2012; Mohr & Webb, 2005; Mohr et al., 2001). Yet, businesses should be mindful of potential gender differences in the perceptions of CSR initiatives.

6 Consumers' Perceptions of CSR: Male Versus Female Perspectives

Researchers concur that males and females often differ in their perspectives, shaped by biological influences and societally constructed gender roles and identities (Bussey & Bandura, 1999; West, 2015). Deeply ingrained perceptions and experiences related to gender traits and roles can shape people's opinions on gender-appropriate actions (Bussey & Bandura, 1999; West, 2015). Sustainable consumption behavior, for example, is strongly tied to feminine rather than masculine traits (Brough et al., 2016; Swim et al., 2011). More specifically, women have always been associated with nurturing and compassionate roles, shaping their stronger inclination toward ethical and sustainable actions (Betz et al., 1989; Borkowski & Ugras, 1992). On the other hand, cultural norms push men toward rigid masculine identities, often sidelining sustainability as this might be seen as too "feminine" (Bosson et al., 2009; Diekman & Goodfriend, 2006). Threats to masculinity may, therefore, deter men from sustainable actions (Brough et al., 2016). Concerns about how one is perceived and accepted by peers might thus influence sustainable consumption behavior to align with traditional gender norms (Bloodhart & Swim, 2020). Here, it is essential to note that both genders may face societal backlash when deviating from norms rooted in societal biases and gendered perceptions (Bashir et al., 2013; Swim et al., 2020).

With the above in mind, research has revealed gender differences among consumers' perceptions of CSR initiatives over several years (Bakewell & Mitchell, 2006; Calabrese et al., 2016; Dunlap, 1983). Earlier studies by Burton and Hegarty (1999) already found that women generally hold higher expectations for an organization's CSR initiatives and display a more profound concern for ethical issues, which, in turn, translates into a stronger propensity to engage in pro-environmental actions. Additionally, researchers found that women tend to be more involved in charitable endeavors, while men often prioritize economic interests within the organization (Ibrahim & Angelidis, 1994; Smith et al., 2001). Women also consistently exhibit higher scores in ethics and social responsibility, with Smith et al. (2001) highlighting their heightened attention to corporate ethical responsibilities. Similarly, Marz et al. (2003) found females had a pronounced social orientation, with Panwar et al. (2010) further revealing differences in male and female perceptions of social responsibility.

More recently, Kahreh et al. (2014) found that women leaned slightly more toward CSR orientation and intentions than men. However, they concluded that gender differences were less significant than in earlier studies. Calabrese et al. (2016) observed that gender differences might be more pronounced in specific dimensions. Notably, the most prominent gender discrepancies they found in CSR perceptions relate to "environmental," "economic," "society," and "labour practice" dimensions (Calabrese et al., 2016). These differences could be attributed to traditional societal roles and expectations for each gender (Windels, 2016). Yet, it would seem that gender variations in CSR perceptions may change and fluctuate over time based on various factors (Calabrese et al., 2016; Fatoki, 2016; Hur et al., 2016). Therefore, more empirical investigation is needed to understand potential gender differences in consumers' perceptions of CSR, particularly in the clothing and textile domain. This understanding is much needed to assess the relevance of gender marketing approaches, particularly in light of recent transitions toward gender-neutral fashion.

7 The Relevance of Gender Marketing in the Clothing Sector

Marketing according to gender differences implies creating and promoting products and advertisements, including those focused on CSR initiatives, aimed at attracting specific genders by leveraging existing societal norms and stereotypes (Putrevu, 2001). The reliance on gender norms has been a prominent strategy in fashion apparel marketing, establishing a link between gender and consumer preferences due to biological differences (Gould & Stern, 1989). For decades, clothing has clearly reflected society's gender norms (Kaiser, 1997). Distinct male and female clothing lines have often perpetuated stereotypes, such as associating pink with girls and blue with boys or females expecting to be immersed in fashion and beauty culture, whereas males are not (Kaiser, 1997; Paoletti, 2012). According to Gould and Stern (1989), the portrayal of gender in advertising frequently reflects existing societal norms. Advertising often reinforces societal norms, establishing a cycle in which fashion marketing and societal gender expectations continuously affect one another. The fashion industry is also a platform where gender roles and ideals of beauty are defined and propagated, such as the ongoing push for more inclusive representation of different body types that do not fit usual socially constructed ideals (Bishop et al., 2018).

Historically, the marketing industry has largely adhered to a binary understanding of gender, which manifested itself in the distinct categorization of products and services designed for men and women. This binary perspective often saw women associated with domestic settings, while men were often portrayed in work-oriented marketing campaigns (Putrevu, 2001). In current society, gender marketing remains a powerful influence in the fashion industry, involving the creation of products, CSR initiatives, and advertisements that strongly appeal to specific gender groups (Avery, 2012). Brands frequently employ gender-focused marketing strategies to delineate niche audiences, cultivating a devoted customer base that resonates with the brand's gender-specific narratives. These narratives can be tailored to resonate with the preferences of these groups through emerging fashion trends, ethical and sustainable marketing involving CSR campaigns, and influencer marketing and partnerships (Rathore, 2019). Luxury brands, mainly, have carefully employed gender-focused marketing to cultivate aspirational images of masculinity or femininity (Stokburger-Sauer & Teichmann, 2013). The goal is to elicit feelings of affluence, influence, and sophistication that harmonize with traditional gender-based narratives, demonstrating a proficient grasp of gender psychology and its impact on consumer behavior (Kapferer & Bastien, 2009).

The fashion industry is, however, rapidly evolving. In response to shifting perceptions, designers and brands are now incrementally introducing elements that contest the established norms of gender-specific dress. Because gender marketing is often deeply anchored in historical and traditional depictions of gender roles, it has been criticized for reinforcing gender stereotypes and restricting individual expression, utilizing stereotypical imagery (Buentello et al., 2022) and conventional gender-specific preferences to strengthen brand identity and increase sales (Grohmann, 2009). As societal perspectives on gender evolve, the fashion industry is experiencing a paradigm shift, moving from a binary approach in gender marketing to a strategy that is more inclusive and gender-neutral. With the introduction of gender-neutral fashion, the transformation in fashion aesthetics represents a move toward greater inclusivity, embracing a diverse portrayal of gender through clothing, reflecting the ongoing advancements in societal norms surrounding gender (Dai et al., 2022).

8 The Shift Toward Gender-Neutral Fashion

The shift to gender-neutral fashion aligns with the growing awareness and acknowledgment of gender-fluid and nonbinary identities, particularly within younger consumer demographics (Britt, 2019). While gender-neutral fashion may appear to be a modern trend, it has historical origins. For example, in the 1920s, Coco Chanel pioneered the introduction of trousers for women, marking a notable shift from conventional female attire (Steele, 1997), thus moving toward a more fluid and androgynous silhouette. The 1960s and 70 s witnessed the unisex trend, marked by androgynous fashion choices, as a significant milestone. During this period, individuals of both genders defied societal expectations and explored fashion choices that blurred traditional gender lines (Chrisman-Campbell, 2015). According to some, Rudi Gernreich pioneered the idea of unisex clothing in 1968, crafting pieces such as trousers and tops meant for both genders (Paoletti, 2015). Yet, when observing these designs closely, it becomes evident that they were primarily tailored for ultra-thin female models, like Twiggy and Jean Shrimpton, during the 1960s, suggesting the clothing catered to women and not men with more masculine body types (Paoletti, 2015). Since the 1990s, major fashion houses and commercial retailers began introducing collections that challenge gender binaries. This trend intensified in the twenty-first century and is not merely a fashion statement. High fashion brands and mainstream retailers amplify this approach, signifying broader societal shifts toward accepting and recognizing diverse gender identities and expressions (Crane, 2012). Today, "gender neutral" is the preferred term over "unisex," denoting clothes designed for all genders (Morgan, 2019).

A significant benefit of this shift toward gender-neutral clothing is the opportunity to access a broader market, including individuals throughout the gender spectrum, thereby fostering a more inclusive and diverse customer base (Lin et al., 2022). For manufacturers of garments, creating gender-neutral clothing is more cost-effective and profitable, as it enables men and women to share the same clothing. Removing the distinction between men's and women's clothing and operating through one production line should inevitably cut costs from research and development to the supply chain. Therefore, producing gender-neutral attire offers manufacturers a lucrative opportunity (Song, 2023). Gender-neutral fashion may also contribute to companies' CSR initiatives in the environmental sphere by, for example, reducing the amount of raw materials and energy needed to create garments, ultimately leading to reduced levels of textile waste (Brydges et al., 2022).

However, shifting toward gender-neutral marketing comes with its own set of challenges. Brands must traverse this territory with genuineness and sensitivity to avoid cases of misrepresentation. This calls for a comprehensive revamping of various aspects, such as product design, supply chain management, and store layouts, to cultivate an inclusive shopping experience (Buentello et al., 2022). Chrisman-Campbell (2015) points out that while the intent behind gender-neutral clothing was to reduce gender distinctions, it often inadvertently emphasized them instead: while the unisex trend might have added masculine elements to women's clothing, it didn't strip them of their femininity. Therefore, brands exploring gender-neutral marketing must be ready for a multidimensional transformation. While these hold promise, they demand a careful and thoughtful approach to effectively connect with the target audience, avoiding the traps of superficiality or misrepresentation, particularly in emerging and developing contexts (Ratten, 2017).

A diverse amalgamation of cultural, historical, and religious contexts has profoundly shaped the fashion landscape in emerging markets. These elements have historically prescribed specific gender norms manifested through attire. In many cultures, where traditions and customs are deeply ingrained, apparel has notably demarcated gender identities, and the conventional clothing designated for women and men has historically been instrumental in outlining gender norms (Kaiser,

1997). Amidst the tide of globalization and heightened exposure to Western fashion trends through diverse platforms and strategies employed by international fashion brands, the formerly stringent boundaries delineating gender norms are progressively becoming more flexible. Urbanization is a critical catalyst in this transformation, fostering a blend of traditional and contemporary fashion narratives, thus exploring a broader and more diverse range of gender expressions within the fashion sector (Hansen, 2004). Nonetheless, these norms can vary, often influenced by religion or marital status (Hansen, 2004; Tarlo, 1996). For these reasons, future research should more diligently address gender influences in the clothing and textile domain, specifically in light of current shifts toward gender neutrality and its potential impact on gender marketing and CSR initiatives.

9 Conceptual Framework and Future Research Recommendations

For many years, research has indicated a growing customer awareness surrounding corporate social and environmental responsibilities (Brunk & Blümelhuber, 2011). However, not every customer shares the same level of interest or knowledge in responsible business practices (Avis et al., 2022). Gender, education, age, ethnicity, income, culture, and nationality may also shape customers' perspectives on a company's responsible behavior (Çarkoğlu & Kentmen-Çin, 2015; Luthar et al., 1997). While these findings may not be universal across all product types or market segments (Auger et al., 2010; Peloza & Shang, 2011), almost all internationally recognized companies aim to brand themselves as socially responsible and environmentally friendly entities, including the clothing sector.

The conceptual framework depicted in Fig. 1 illustrates the various relationships between CSR dimensions and concepts derived from insights established by Öberseder et al. (2014) and Alvarado-Herrera et al. (2017). In adopting a TBL approach, clothing manufacturers and retailers may impact consumers' perceptions of CSR based on the three core dimensions Alvarado-Herrera et al. (2017) identified: economic, social, and environmental. The original theoretical basis proposed by Öberseder et al. (2014) was founded upon seven distinguishable domains of CSR that focus on a company's stakeholders, including the community, employees, societal, customers, suppliers, shareholders, and the environment. Consumer perceptions of CSR may, in turn, contribute to clothing purchase intentions and the accompanying support of particular clothing brands (Dickson, 2000; Diddi & Niehm, 2017).

Gender has been a focal point in studying its influence on consumers' perceptions of CSR, especially compared to other demographic elements. Although numerous studies have delved into gender perspectives and perceptions of CSR, the findings remain somewhat inconsistent, with scholars differing on the importance of gender disparities (Calabrese et al., 2016; Collins, 2000; McCabe et al., 2006;

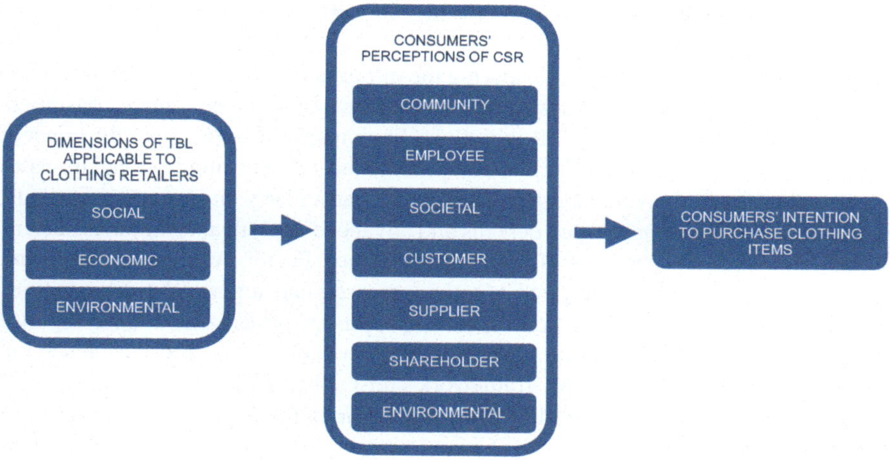

Fig. 1 The adapted conceptual framework derived from theoretical foundations of TBL dimensions by Alvarado-Herrera et al. (2017), and Consumer Perceptions of CSR by Öberseder et al. (2014)

Singhapakdi et al., 2001). Butt (2016) emphasizes the importance of continued research in this area, particularly in emerging economies where a limited understanding of gender differences in consumer behavior exists. Fatoki (2016) identified significant differences between male and female students' perceptions of CSR in a study across three universities in the South African emerging market context. Similarly, Hur et al. (2016) found notable gender disparities in CSR perceptions among consumers in South Korea. The question, however, remains whether these gender differences prevail in the clothing sector, which has seen substantial shifts toward gender neutrality over the past few years.

In emerging markets, deeply ingrained traditions, religious doctrines, and societal norms shape gender roles and perceptions, influencing everyday consumption and purchasing behaviors (Jayachandran, 2015). Gender roles often influence sustainable consumption behaviors, with stereotypes shaping male and female engagement differently (Dahl et al., 2013). Various studies show that women typically engage more in sustainable consumption activities than men (Hunter et al., 2004; Zelezny et al., 2000). However, women's sustainable consumption extends to purchasing eco-friendly products, like organic food and sustainable clothing. In contrast, men often embrace sustainable activities tied to traditional male roles, such as installing energy-efficient technologies (Dahl et al., 2013). Still, males tend to avoid sustainable consumption behaviors perceived as feminine, such as recycling or using reusable bags, suggesting societal stereotypes might hinder men more than women in adopting certain sustainable practices (Brough et al., 2016; Johnson et al., 2013). Lastly, it is essential to consider that income disparity and intersections of gender with social class can significantly impact these consumption patterns, with affluent individuals consuming more and gendered consumption disparities more pronounced among the economically disadvantaged (Axsen et al., 2012; Johnsson-Latham, 2007; Kurz et al., 2015).

Against this backdrop and considering the importance of CSR in the clothing and textile industry, future research should explore whether male and female consumers differ in their perceptions of CSR initiatives. Such research could apply the conceptual framework depicted in Fig. 1 among male and female sample populations to determine whether potential differences exist, particularly in emerging markets. Utilizing cross-sectional survey designs and advanced analytical techniques such as Exploratory Factor Analysis (EFA), Confirmatory Factor Analysis (CFA), and Structural Equation Modeling (SEM), this research can unpack the complexities surrounding male and female consumers' perceptions, prioritization, and response to clothing retailers' CSR efforts. These insights will not only enrich the academic discourse surrounding the TBL framework and gender differences in consumers' perceptions of CSR but also guide businesses operating in culturally diverse settings.

Future research could also benefit from interpreting results based on gender theories (such as social identity theory and gender schema theory) to unveil the underlying biases and schemas that shape and often constrain routes to sustainable consumption, offering valuable insights for enhancing CSR initiatives in the clothing industry. Applying these theoretical frameworks to CSR strategies could facilitate creating and refining approaches that align with consumers' varied gender-specific demands and preferences, promoting a comprehensive and gender-sensitive sustainability narrative in the industry. Hence, integrating gender theories into strategic CSR efforts is a potential pathway that holds substantial promise in driving transformative changes toward sustainable consumption in the clothing and textile sector.

10 Applying Gender Theories to Interpret Gender Influences

Employing gender theories to interpret results is instrumental in appreciating the subtle ways in which gender influences consumer attitudes and behaviors. This refined understanding is crucial for creating CSR strategies that resonate with a broad spectrum of consumers, cultivating an environment that both confronts and supports diverse gender identities. A revolutionary approach to CSR, enriched with profound insights from gender theories, carries the transformative power to ignite a shift toward sustainable practices that resonate with purpose and significance. The clothing and textile industry could benefit immensely from incorporating perspectives sensitive to gender nuances within their strategic paradigms.

Although many gender theories have emerged over the past few years, the selection of social identity theory and gender schema theory is rooted in their comprehensive examination of gender relations and their consequential impact on individual choices, especially in the clothing and textile industry. Understanding the intricate processes by which individuals categorize themselves and conform to societal gender expectations is vital for comprehending the multifaceted consumer environment. Such insights equip brands with the knowledge to adeptly maneuver and harness the intricate nexus of gender identities and societal standards, thereby promoting sustainability within the clothing and textile sector. When investigating the influence of gender theories on consumer patterns in fashion, it is crucial to employ models that deeply explore the subtle intricacies of gender interplay and its overlap with societal expectations. The social identity theory and the gender schema theory stand out, offering rich, analytical lenses that allow for a deep understanding of the complex interplay between gender, identity, and possible consumption patterns.

10.1 Social Identity Theory

Social identity theory is a foundational pillar in social psychology, offering insights into how individuals perceive, establish relations with, and are shaped through their affiliations to different social groups (Tajfel & Turner, 1978, 2004). These affiliations are pivotal in shaping individuals' self-concept, guiding behavior, and influencing interpersonal interactions. Social identity theory emphasizes the deep-seated impact of group classifications on intergroup dynamics (Tajfel & Turner, 2004). This theory proposes that individuals draw a significant portion of their identity from group affiliations. These affiliations often cause marked bias by favoring one's group—the ingroup—and can result in potential discrimination against other groups or outgroups (Brewer, 1991; Brewer & Kramer, 1985; Tajfel & Turner, 1978).

The clothing and textile sector provides a palpable domain where the interplay of social identity theory and gender could manifest. Historically, fashion has been a formidable tool to enable individuals to express themselves visibly or, at times, challenge their gender identities. For instance, specific clothing items underwent significant sociocultural transitions (Akdemir, 2018). As such, initially symbolizing masculinity, women's widespread adoption of pants during the 1960s symbolized burgeoning gender equality (Crane, 2012; Morgan, 2019; Paoletti, 2015).

The contemporary fashion landscape is teeming with innovative expressions that defy traditional gender norms. Modern trends featuring androgynous style designs symbolize an evolving conversation around gender fluidity and the multifaceted spectrum of gender identities (Britt, 2019). These decisions aren't merely about aesthetic preferences; they echo the more extensive societal dialogue, challenging rigid gender constructs and advocating for a more inclusive understanding of identity. The social identity theory thus emphasizes the profound implications of an individual's affiliations with social groups. These affiliations are pivotal in decision-making processes (Tajfel & Turner, 1978, 2004) and can subsequently impact consumption patterns. For businesses, recognizing the implications of social identity theory means understanding that consumers don't just buy products or services; they buy into identities, which may include those related to CSR initiatives.

10.2 Gender Schema Theory

Gender schema theory, introduced by Bem (1981), posits that children's cognition and actions are shaped by active schemas, which influence their gender identity and self-perceptions. The theory underscores the process over content, highlighting how children, through gender-based schematic processing, adopt characteristics deemed appropriate for their respective genders. This results from intertwining one's self-concept with the gender schema, a cognitive structure classifying gender-related beliefs, lifestyles, and sexuality (Bem, 1981; West, 2015). Such attributes, shaped by societal norms, determine what is appropriate for each gender.

Sex-typing emerges partly from this schematic processing, where there's an innate tendency to process information rooted in gender associations (Bem, 1981). Consequently, children learn the more profound relevance of specific gender attributes. Sex-typed individuals are not differentiated by their degree of masculinity or femininity but by how they mold their self-concept and behavior around gender. Society provides a vast network of gender associations that can evolve into different cognitive schemata while simultaneously stressing the profound importance of the gender divide in most life facets (Bem, 1981). As children assimilate various associative networks over time, gender, due to its perceived importance, stands out as a predominant and easily accessible schema (Bem, 1981). Gender schema theory thus emphasizes the intricate mental maps we develop about gender from a young age (Bem, 1981). It is more than just recognizing what is "male" and "female." It is about how deeply these notions influence choices throughout one's life.

In the context of sustainable fashion, this means that consumer preferences might be significantly swayed by how well products align with gender stereotypes. For instance, if sustainable practices and CSR initiatives are generally associated with a particular gender, it could influence the buying behaviors of individuals identifying with that gender more strongly. Harnessing the gender schema theory, CSR endeavors can question or support existing gender schemas to propagate sustainable habits. By aligning sustainable fashion choices with prevailing gender conventions, companies might accomplish a more profound influence, spurring consumers to adopt eco-friendly choices based on their gender identities.

In summary, CSR initiatives targeting sustainable practices in the clothing and textile industry stand to gain insights from the social identity theory and gender schema theory. These theoretical frameworks shed light on the subtle interplay of gender dynamics and their influence on consumer behavior, equipping businesses with a plan to devise CSR strategies that deeply resonate with their audience. When sustainable consumption aligns with femininity, companies can employ social identity theory by strategizing initiatives emphasizing women's sustainability empowerment. This approach appeals to the "ingroup" identity of women. Conversely, addressing men's reluctance toward sustainable choices due to potential masculinity threats, companies are positioned to engineer campaigns challenging such perceptions, associating sustainability with universal virtues over gendered associations. Additionally, tapping into the social identity theory can further support CSR campaigns by endorsing fashion brands that reflect or challenge the gender identities of their clientele, fostering a more inclusive, sustainable fashion ecosystem.

11 Conclusion

Recent years have seen society becoming increasingly aware of and concerned with social and environmental matters, as well as within the clothing and textile industry. Consequently, companies face mounting pressure to align their operations with societal expectations. The foundational concept of CSR and the accompanying TBL approach asserts that companies should evaluate and disclose their achievements not merely in financial terms (profit) but also through their social (people) and environmental (planet) contributions. Numerous businesses worldwide have thus woven TBL values into their daily operations. The TBL framework alters business strategies and holds varying resonance across gender distinctions. Several studies have shown differences between genders in their perceptions of environmental and social responsibilities, which subsequently necessitates companies' adaptation and execution of the TBL approach. The complex relationship between gender constructs and sustainability principles in the clothing and textile industry is paramount in evolving CSR initiatives and promoting sustainable consumer practices. Recent trends toward gender-neutral apparel also add new dimensions to the complexity of gender influences on sustainable clothing consumption behavior.

Stakeholders in the clothing and textile supply chain can capitalize on further insights surrounding gender influences in two pivotal ways: firstly, in product development and positioning, companies may either recalibrate product positioning or challenge entrenched gender schemas if sustainable items are gendered. If certain eco-friendly products are labeled typically "feminine," a challenge lies in marketing them to resonate with both genders. Through this lens, businesses can offer products that change stereotypes, endorsing gender-neutral sustainability. Secondly, acknowledging the influence of gender schemas enables corporations to undertake a vital role in societal education. Initiatives that disrupt traditional gender norms help businesses break stereotypes and correspond with CSR values.

Employing theories such as social identity theory and gender schema theory is instrumental in appreciating how gender influences consumer attitudes and behaviors. This refined understanding is crucial for creating CSR strategies that resonate with a broad spectrum of consumers, cultivating an environment that both confronts and supports diverse gender identities. A revolutionary approach to CSR, enriched with profound insights from gender theories, carries the transformative power to ignite a shift toward sustainable practices that resonate with purpose and significance.

References

Aguinis, H., & Glavas, A. (2012). What we know and don't know about corporate social responsibility: A review and research agenda. *Journal of Management, 38*(4), 932–968.

Akdemir, N. (2018). Visible expression of social identity: The clothing and fashion. *Gaziantep University Journal of Social Sciences, 17*(4), 1389–1397.

Alhaddi, H. (2015). Triple bottom line and sustainability: A literature review. *Business and Management Studies, 1*(2), 6–10.

Alvarado-Herrera, A., Bigne, E., Aldas-Manzano, J., & Curras-Perez, R. (2017). A scale for measuring consumer perceptions of corporate social responsibility following the sustainable development paradigm. *Journal of Business Ethics, 140*(2), 243–262.

Auger, P., Devinney, T. M., Louviere, J. J., & Burke, P. F. (2010). The importance of social product attributes in consumer purchasing decisions: A multi-country comparative study. *International Business Review, 19*(2), 140–159.

Avery, J. (2012). Defending the markers of masculinity: Consumer resistance to brand gender-bending. *International Journal of Research in Marketing, 29*(4), 322–336.

Avis, M., Konopka, R., Gregory-Smith, D., & Palakshappa, N. (2022). Disentangling consumers' CSR knowledge types and effects. *Sustainability, 14*(19), 1–20.

Axsen, J., TyreeHageman, J., & Lentz, A. (2012). Lifestyle practices and pro-environmental technology. *Ecological Economics, 82*, 64–74.

Bakewell, C., & Mitchell, V.-W. (2006). Male versus female consumer decision making styles. *Journal of Business Research, 59*(12), 1297–1300.

Barber, E. J. W. (1994). *Womens work: The first 20000 years women cloth and society in early times*. WW Norton & Company.

Bashir, N. Y., Lockwood, P., Chasteen, A. L., Nadolny, D., & Noyes, I. (2013). The ironic impact of activists: Negative stereotypes reduce social change influence. *European Journal of Social Psychology, 43*(7), 614–626.

Bem, S. L. (1981). Gender schema theory: A cognitive account of sex typing. *Psychological Review, 88*(4), 354.

Betz, M., O'Connell, L., & Shepard, J. M. (1989). Gender differences in proclivity for unethical behavior. *Journal of Business Ethics, 8*, 321–324.

Bhatia, N., & Bhatia, S. (2021). Changes in gender stereotypes over time: A computational analysis. *Psychology of Women Quarterly, 45*(1), 106–125.

Bishop, K., Gruys, K., & Evans, M. (2018). Sized out: Women, clothing size, and inequality. *Gender & Society, 32*(2), 180–203.

Bittman, M., England, P., Sayer, L., Folbre, N., & Matheson, G. (2003). When does gender trump money? Bargaining and time in household work. *American Journal of Sociology, 109*(1), 186–214.

Bloodhart, B., & Swim, J. K. (2020). Sustainability and consumption: What's gender got to do with it? *Journal of Social Issues, 76*(1), 101–113.

Blowfield, M., & Murray, A. (2014). *Corporate responsibility*. Oxford University Press.

Borkowski, S. C., & Ugras, Y. J. (1992). The ethical attitudes of students as a function of age, sex and experience. *Journal of Business Ethics, 11*, 961–979.

Bosson, J. K., Vandello, J. A., Burnaford, R. M., Weaver, J. R., & Arzu Wasti, S. (2009). Precarious manhood and displays of physical aggression. *Personality and Social Psychology Bulletin, 35*(5), 623–634.

Bowen, H. R. (1953). *Social responsibilities of the businessman*. Harper.

Brewer, M. B. (1991). The social self: On being the same and different at the same time. *Personality and Social Psychology Bulletin, 17*(5), 475–482.

Brewer, M. B., & Kramer, R. M. (1985). The psychology of intergroup attitudes and behavior. *Annual Review of Psychology, 36*(1), 219–243.

Britt, M. (2019). *Defining Gender Neutral Fashion*. Unpublished work.

Brough, A. R., Wilkie, J. E., Ma, J., Isaac, M. S., & Gal, D. (2016). Is eco-friendly unmanly? The green-feminine stereotype and its effect on sustainable consumption. *Journal of Consumer Research, 43*(4), 567–582.

Brunk, K. H., & Blümelhuber, C. (2011). One strike and you're out: Qualitative insights into the formation of consumers' ethical company or brand perceptions. *Journal of Business Research, 64*(2), 134–141.

Brydges, T., Henninger, C. E., Barbu, L., & Lupu, R. (2022). Designing for longevity and neutrality: Investigating how the Swedish children's clothing industry implements circular economy principles. *Fashion Practice, 15*(3), 424–446.

Buentello, D. A., Bardey, A., & Rogaten, J. (2022). Beyond the gender dichotomy in fashion: Exploring the factors involved in cross-sexual fashion consumer behaviour for cisgender women. *Journal of Fashion Marketing and Management: An International Journal, 27*(4), 697–709.

Burton, B. K., & Hegarty, W. H. (1999). Some determinants of student corporate social responsibility orientation. *Business & Society, 38*(2), 188–205.

Bussey, K., & Bandura, A. (1999). Social cognitive theory of gender development and differentiation. *Psychological Review, 106*(4), 676.

Butler, J. (2011). *Gender trouble: Feminism and the subversion of identity*. Routledge.

Butt, I. (2016). Corporate social responsibility and consumer buying behavior in emerging market: A mixed method study. *International Journal of Business and Management, 11*(7), 211–222.

Calabrese, A., Costa, R., & Rosati, F. (2016). Gender differences in customer expectations and perceptions of corporate social responsibility. *Journal of Cleaner Production, 116*, 135–149.

Cannon, T. B. S. (1992). *Corporate responsibility*. Pitman.

Çarkoğlu, A., & Kentmen-Çin, Ç. (2015). Economic development, environmental justice, and pro-environmental behavior. *Environmental Politics, 24*(4), 575–597.

Carroll, A. B. (1991). The pyramid of corporate social responsibility: Toward the moral management of organizational stakeholders. *Business Horizons, 34*(4), 39–48.

Chadwick, W., & Chadwick, W. (1990). *Women, art, and society*. Thames and Hudson.

Chrisman-Campbell, K. (2015). *A brief history of unisex fashion.* [Online] https://www.theatlantic.com/entertainment/archive/2015/04/when-unisex-was-the-new-black/390168/#. Accessed 20 Mar 2023

Collins, D. (2000). The quest to improve the human condition: The first 1 500 articles published in journal of business ethics. *Journal of Business Ethics, 26*(1), 1–73.

Crane, D. (2012). *Fashion and its social agendas: Class, gender, and identity in clothing*. University of Chicago Press.

Dahl, J., Vescio, T., Swim, J., & Johnson, S. (2013). *Masculinity and pro-environmental engagement.* Unpublished manuscript.

Dai, H., Xia, T., & Zhao, S. (2022). Unisex style: The history, current situation and future. *Advances in Social Science, Education and Humanities Research, 664*, 1127–1133.

Das, M. (2022). Empowering women through CSR initiatives. *IUP Journal of Management Research, 21*(4), 32–43.

Davidson, D. J., & Freudenburg, W. R. (1996). Gender and environmental risk concerns: A review and analysis of available research. *Environment and Behavior, 28*(3), 302–339.

Dawkins, C., & Ngunjiri, F. W. (2008). Corporate social responsibility reporting in South Africa: A descriptive and comparative analysis. *The Journal of Business Communication (1973), 45*(3), 286–307.

De Vries, J. (1994). The industrial revolution and the industrious revolution. *The Journal of Economic History, 54*(2), 249–270.

Dickson, M. A. (2000). Personal values, beliefs, knowledge, and attitudes relating to intentions to purchase apparel from socially responsible businesses. *Clothing and Textiles Research Journal, 18*(1), 19–30.

Dickson, M. A., & Eckman, M. (2006). Social responsibility: The concept as defined by apparel and textile scholars. *Clothing and Textiles Research Journal, 24*(3), 178–191.

Diddi, S., & Niehm, L. S. (2017). Exploring the role of values and norms towards consumers' intentions to patronize retail apparel brands engaged in corporate social responsibility (CSR). *Fashion and Textiles, 4*(1), 1–20.

Diekman, A. B., & Goodfriend, W. (2006). Rolling with the changes: A role congruity perspective on gender norms. *Psychology of Women Quarterly, 30*(4), 369–383.

Du, S., Bhattacharya, C. B., & Sen, S. (2011). Corporate social responsibility and competitive advantage: Overcoming the trust barrier. *Management Science, 57*(9), 1528–1545.

Dunlap, R. E. (1983). Male-female differences in concern for environmental quality. *International Journal of Women's Studies, 6*(4), 291–301.

Eells, R., & Walton, C. C. (1961). *Conceptual foundations of business*. R.D. Irwin.

Eisenberg, N., & Lennon, R. (1983). Sex differences in empathy and related capacities. *Psychological Bulletin, 94*(1), 100.

Eisend, M. (2010). A meta-analysis of gender roles in advertising. *Journal of the Academy of Marketing Science, 38*, 418–440.

Elkington, J. (1997). *Cannibals with forks: The triple bottom line of twenty-first century business*. Capstone.

Fatoki, O. (2016). Gender and the perception of corporate social responsibility by university students in South Africa. *Gender and Behaviour, 14*(3), 7574–7588.

Fausto-Sterling, A. (2000). *Sexing the body: Gender politics and the construction of sexuality*. Basic Books.

Fulton, K., & Lee, S.-E. (2013). Assessing sustainable initiatives of apparel retailers on the internet. *Journal of Fashion Marketing and Management: An International Journal, 17*(3), 353–366.

Goel, P. (2010). Triple bottom line reporting: An analytical approach for corporate sustainability. *Journal of Finance, Accounting & Management, 1*(1), 27–42.

Gould, S. J., & Stern, B. B. (1989). Gender schema and fashion consciousness. *Psychology & Marketing, 6*(2), 129–145.

Grau, S. L., & Zotos, Y. C. (2016). Gender stereotypes in advertising: A review of current research. *International Journal of Advertising, 35*(5), 761–770.

Grohmann, B. (2009). Gender dimensions of brand personality. *Journal of Marketing Research, 46*(1), 105–119.

Gruber, V., Schlegelmilch, B. B., & Houston, M. J. (2014). Inferential evaluations of sustainability attributes: Exploring how consumers imply product information. *Psychology & Marketing, 31*(6), 440–450.

Hansen, K. T. (2004). The world in dress: Anthropological perspectives on clothing, fashion, and culture. *Annual Review of Anthropology, 33*, 369–392.

Henry, L. A., Buyl, T., & Jansen, R. J. (2019). Leading corporate sustainability: The role of top management team composition for triple bottom line performance. *Business Strategy and the Environment, 28*(1), 173–184.

Hilson, G. (2012). Corporate social responsibility in the extractive industries: Experiences from developing countries. *Resources Policy, 37*(2), 131–137.

Hobson, J. (2013). To die for? The health and safety of fast fashion. *Occupational Medicine, 63*(5), 317–319.

Hunter, L. M., Hatch, A., & Johnson, A. (2004). Cross-national gender variation in environmental behaviors. *Social Science Quarterly, 85*(3), 677–694.

Hur, W. M., Kim, H., & Jang, J. H. (2016). The role of gender differences in the impact of CSR perceptions on corporate marketing outcomes. *Corporate Social Responsibility and Environmental Management, 23*(6), 345–357.

Ibrahim, N. A., & Angelidis, J. P. (1994). Effect of board members' gender on corporate social responsiveness orientation. *Journal of Applied Business Research, 10*(1), 35.

Jägel, T., Keeling, K., Reppel, A., & Gruber, T. (2012). Individual values and motivational complexities in ethical clothing consumption: A means-end approach. *Journal of Marketing Management, 28*(3–4), 373–396.

Jamali, D. (2006). Insights into triple bottom line integration from a learning organization perspective. *Business Process Management Journal, 12*(6), 809–821.

Jansen van Rensburg, M. (2016). Chapter 5: Corporate social responsibility. In B. Erasmus, J. W. Strydom, & S. Rudansky-Kloppers (Eds.), *Introduction to business management* (pp. 146–176). Oxford University Press Southern Africa.

Jayachandran, S. (2015). The roots of gender inequality in developing countries. *Annual Review of Economics, 7*(1), 63–88.

Johnson, S., Dahl, J., Swim, J., & Vescio, T. (2013). The gender-stereotypic nature of climate change positions. *Society for Personality and Social Psychology*. https://doi.org/10.1037/e514472015-187

Johnsson-Latham, G. (2007). A study on gender equality as a prerequisite for sustainable development. In *Report to the Environment Advisory Council*. 2007:2. Environment Advisory Council, Ministry Of The Environment.

Kahreh, M. S., Babania, A., Tive, M., & Mirmehdi, S. M. (2014). An examination to effects of gender differences on the corporate social responsibility (CSR). *Procedia-Social and Behavioral Sciences, 109*, 664–668.

Kaiser, S. B. (1997). *The social psychology of clothing: Symbolic appearances in context*. Fairchild.

Kapferer, J.-N., & Bastien, V. (2009). The specificity of luxury management: Turning marketing upside down. *Journal of Brand Management, 16*(5–6), 311–322.

Kotler, P., & Lee, N. (2005). *Corporate social responsibility: Doing the most good for your company and your cause*. Wiley.

Kurz, T., Gardner, B., Verplanken, B., & Abraham, C. (2015). Habitual behaviors or patterns of practice? Explaining and changing repetitive climate-relevant actions. *Wiley Interdisciplinary Reviews: Climate Change, 6*(1), 113–128.

Lichtenstein, D. R., Drumwright, M. E., & Braig, B. M. (2004). The effect of corporate social responsibility on customer donations to corporate-supported nonprofits. *Journal of Marketing, 68*(4), 16–32.

Lillywhite, S., & Diviney, E. (2007). *Ethical threads: Corporate social responsibility in the Australian garment industry*. Laurence, B.O.S.

Lim, L. Y. (1990). Women's work in export factories: The politics of a cause. In *Business, Government and Labor: Essays on Economic Development in Singapore and Southeast Asia* (pp. 217–240). World Scientific Publishing Company.

Lin, L., Liu, J., Yang, M., & Zhang, Z. (2022). The Prospect of non-sexist clothing in the future clothing market. *Journal of Education, Humanities and Social Sciences, 5*, 1–5.

Luthar, H. K., DiBattista, R. A., & Gautschi, T. (1997). Perception of what the ethical climate is and what it should be: The role of gender, academic status, and ethical education. *Journal of Business Ethics, 16*, 205–217.

Marz, J. W., Powers, T. L., & Queisser, T. (2003). Corporate and individual influences on manager' social orientation. *Journal of Business Ethics, 46*, 1–11.

McCabe, A. C., Ingram, R., & Dato-On, M. C. (2006). The business of ethics and gender. *Journal of Business Ethics, 64*(2), 101–116.

Mersham, G. M., & Skinner, C. (2016). South Africa's bold and unique experiment in CSR practice. *Society and Business Review, 11*(2), 110–129.

Mohr, L. A., & Webb, D. J. (2005). The effects of corporate social responsibility and price on consumer responses. *Journal of Consumer Affairs, 39*(1), 121–147.

Mohr, L. A., Webb, D. J., & Harris, K. E. (2001). Do consumers expect companies to be socially responsible? The impact of corporate social responsibility on buying behavior. *Journal of Consumer Affairs, 35*(1), 45–72.

Moir, L. (2001). What do we mean by corporate social responsibility? *Corporate Governance: The international Journal of Business in Society, 1*(2), 16–22.

Molly, M., & Larner, W. (2010). Who needs cultural intermediaries indeed. *Journal of Cultural Economy, 3*(3), 361–377.

Morgan, A. (2019). *Why the terms unisex and gender neutral are not fit for purpose in contemporary clothing and fashion design*. Cardiff Metropolitan University.

Öberseder, M., Schlegelmilch, B. B., Murphy, P. E., & Gruber, V. (2014). Consumers' perceptions of corporate social responsibility: Scale development and validation. *Journal of Business Ethics, 124*(1), 101–115.

Panwar, R., Hansen, E., & Anderson, R. (2010). Students' perceptions regarding CSR success of the US forest products industry. *Social Responsibility Journal, 6*(1), 18–32.

Paoletti, J. (2012). *Pink and blue: Telling the boys from the girls in America*. Indiana University Press.

Paoletti, J. (2015). *Sex and unisex: Fashion, feminism, and the sexual revolution*. Indiana University Press.

Park, H., & Kim, Y.-K. (2016). An empirical test of the triple bottom line of customer-centric sustainability: The case of fast fashion. *Fashion and Textiles, 3*(1), 1–18.

Peloza, J., & Shang, J. (2011). How can corporate social responsibility activities create value for stakeholders? A systematic review. *Journal of the Academy of Marketing Science, 39*, 117–135.

Pérez, R. C. (2009). Effects of perceived identity based on corporate social responsibility: The role of consumer identification with the company. *Corporate Reputation Review, 12*(2), 177–191.

Perumall, A. (2023). *South Africa greylisted: The financial action task force plenary adopts report on South African anti-money laundering and counter-terrorist financing measures*. [Online] https://insightplus.bakermckenzie.com/bm/banking-finance_1/south-africa-greylisted-the-financial-action-task-force-plenary-adopts-report-on-south-african-anti-money-laundering-and-counter-terrorist-financing-measures. Accessed 20 Mar 2023.

Putrevu, S. (2001). Exploring the origins and information processing differences between men and women: Implications for advertisers. *Academy of Marketing Science Review, 10*(1), 1–14.

Rathore, B. (2019). Chic strategies: Revolutionizing the industry through innovative fashion marketing. *International Journal of New Media Studies: International Peer Reviewed Scholarly Indexed Journal, 6*(2), 23–33.

Ratten, V. (2017). Gender entrepreneurship and global marketing. *Journal of Global Marketing, 30*(3), 114–121.

Ross, R. J. S. (2004). *Slaves to fashion : Poverty and abuse in the new sweatshops*. University of Michigan Press.

Russo, A., & Perrini, F. (2010). Investigating stakeholder theory and social capital: CSR in large firms and SMEs. *Journal of Business Ethics, 91*(2), 207–221.

Ruwanpura, K. N. (2011). Women workers in the apparel sector: A three decade (r)-evolution of feminist contributions? *Progress in Development Studies, 11*(3), 197–209.

Ryan, B. 2007. Sex and Gender. In *The Blackwell Encyclopedia of Sociology, G. Ritzer (Ed)* (pp. 4204–4206). Blackwell Publishing.

Schwartz, M. S., & Carroll, A. B. (2003). Corporate social responsibility: A three-domain approach. *Business Ethics Quarterly, 13*(4), 503–530.

Sen, A. (2001). The many faces of gender inequality. *New Republic, 18*(22), 466–477.

Shen, B., Wang, Y., Lo, C. K., & Shum, M. (2012). The impact of ethical fashion on consumer purchase behavior. *Journal of Fashion Marketing and Management: An International Journal, 16*(2), 234–245.

Simionescu, L. N. (2015). *The gender differences effects on corporate social responsibility (CSR): Vol. 3 annals of Constantin, economy series*. Brancusi University of Targu-Jiu.

Singhapakdi, A., Karande, K., Rao, C., & Vitell, S. J. (2001). How important are ethics and social responsibility?-a multinational study of marketing professionals. *European Journal of Marketing, 35*(1/2), 133–153.

Smith, W. J., Wokutch, R. E., Harrington, K. V., & Dennis, B. S. (2001). An examination of the influence of diversity and stakeholder role on corporate social orientation. *Business & Society, 40*(3), 266–294.

Song, J. (2023). The development of unisex clothing. *Journal of Education, Humanities and Social Sciences, 8*, 2543–2547.

Statistics South Africa. (2022). *Quarterly Labour Force Survey (QLFS) Q2 2022*. [Online] https://www.statssa.gov.za/?p=15685. Accessed 10 Oct 2023.

Steele, V. (1997). *Fifty years of fashion : New look to now*. Yale University Press.
Stokburger-Sauer, N. E., & Teichmann, K. (2013). Is luxury just a female thing? The role of gender in luxury brand consumption. *Journal of Business Research, 66*(7), 889–896.
Swim, J. K., Clayton, S., & Howard, G. S. (2011). Human behavioral contributions to climate change: Psychological and contextual drivers. *American Psychologist, 66*(4), 251.
Swim, J. K., Gillis, A. J., & Hamaty, K. J. (2020). Gender bending and gender conformity: The social consequences of engaging in feminine and masculine pro-environmental behaviors. *Sex Roles, 82*(5), 363–385.
Tajfel, H., & Turner, J. C. (1978). *Intergroup behavior. Introducing social psychology, 401*, 466.
Tajfel, H., & Turner, J. C. (2004). The social identity theory of intergroup behavior. In *Political psychology* (pp. 276–293). Psychology Press.
Tarlo, E. (1996). *Clothing matters: Dress and identity in India*. University of Chicago Press.
Thawre, V., & Singh, D. (2021). Gender based violence. In *Indian Economy: Deceleration and Unsustainability* (pp.307–308). Devyani Publishers and Distributors.
The World Bank Group. (2022). *Population % South Africa*. [Online]. https://data.worldbank.org/indicator/SP.POP.TOTL.FE.ZS?locations=ZA. Accessed 07 Oct 2023.
Tsarenko, Y., Ferraro, C., Sands, S., & McLeod, C. (2013). Environmentally conscious consumption: The role of retailers and peers as external influences. *Journal of Retailing and Consumer Services, 20*(3), 302–310.
United Nations. (2008). *Violence against women: A brief overview of the United Nations and violence against women*. [Online] https://www.un.org/womenwatch/daw/vaw/v-overview.htm. Accessed 17 Jan 2024.
United Nations. (2015). *Transforming our world: The 2030 agenda for sustainable development*. [Online] https://sdgs.un.org/2030agenda. Accessed 18 Sept 2023.
Vicente-Molina, M. A., Fernández-Sáinz, A., & Izagirre-Olaizola, J. (2013). Environmental knowledge and other variables affecting pro-environmental behaviour: Comparison of university students from emerging and advanced countries. *Journal of Cleaner Production, 61*, 130–138.
Vijeyarasa, R., & Liu, M. (2022). Fast fashion for 2030: Using the pattern of the sustainable development goals (SDGs) to cut a more gender-just fashion sector. *Business and Human Rights Journal, 7*(1), 45–66.
West, A. (2015). A brief review of cognitive theories in gender development. *Behavioural Sciences Undergraduate Journal, 2*(1), 59–66.
White, C. L., Nielsen, A. E., & Valentini, C. (2017). CSR research in the apparel industry: A quantitative and qualitative review of existing literature. *Corporate Social Responsibility and Environmental Management, 24*(5), 382–394.
Windels, K. (2016). Stereotypical or just typical: How do US practitioners view the role and function of gender stereotypes in advertisements? *International Journal of Advertising, 35*(5), 864–887.
Wood, D. J. (1991). *Corporate social performance revisited*. Wissenschaftszentrum Berlin für Sozialforschung.
Wood, W., & Eagly, A. H. (2012). Biosocial construction of sex differences and similarities in behavior. In *Advances in experimental social psychology* (pp. 55–123). Elsevier Inc.
Zelezny, L. C., Chua, P. P., & Aldrich, C. (2000). New ways of thinking about environmentalism: Elaborating on gender differences in environmentalism. *Journal of Social Issues, 56*(3), 443–457.

Advancing Integration of CSR and Social Life Cycle Assessment in the Textile and Fashion Industry

Anubhuti Bhatnagar and Kirsi Niinimäki

Abstract This chapter shows how Social Life Cycle Assessment (S-LCA) could be integrated with Corporate Social Responsibility (CSR) to monitor the challenges emerging from the textile industry's transition towards a circular economy (CE). While CSR and S-LCA share a common objective of addressing societal challenges resulting from industrial practices, they are distinct tools. CSR outlines a company's voluntary commitment to mitigate its societal, financial, and environmental impacts. S-LCA enables the identification, measurement, and evaluation of social hotspots throughout a product's life cycle. Embedding CSR with S-LCA is vital because the CE transition may alter material supply chains, which could significantly impact stakeholders such as suppliers, employees or workers, local communities, and consumers.

Keywords Social hotspots · Product Life Cycle Assessment (LCA) · Circular economy (CE) · Sustainable textile strategy · Socially responsible practices · Stakeholders · Fast fashion

1 Sustainability and circular economy in the textile sector

The textile industry includes activities like the transformation of raw materials (synthetic or natural) into fibers, yarns, and fabrics, which are then turned into clothing, upholstery, industrial filters, and medical textiles. Over the past two decades, the way textiles, especially clothes, are produced and consumed has changed significantly. In 2015, the fast fashion industry had annual earnings of 1.8 trillion US dollars, which is likely to reach up to 2.1 trillion US dollars by 2025 (Thorisdottir & Johannsdottir, 2020).

A. Bhatnagar (✉) · K. Niinimäki
Department of Design, Aalto University, Espoo, Finland
e-mail: anubhuti.bhatnagar@aalto.fi

S. S. Muthu (ed.), *Corporate Social Responsibility in Textiles and Fashion*,
Sustainable Textiles: Production, Processing, Manufacturing & Chemistry,
https://doi.org/10.1007/978-3-031-61099-8_5

The rapid growth in demand for clothes at low prices has pushed this industry to become a global enterprise. For instance, countries in Asia like China, Bangladesh, India, Pakistan, Vietnam, Malaysia, Indonesia, and Hong Kong accounted for 54% of the total global clothing exports in 2021 (World Trade Organization, 2022). Whereas countries in the European Union, Canada, United States of America, United Kingdom, Switzerland, and Australia accounted for ~62% of the clothing imported (World Trade Organization, 2022). These long textile supply chains are often accused of low levels of transparency regarding the potential for environmental damage in production or waste management, and societal harm through unethical labor practices (Bubicz et al., 2021). Further, the scale of production is also driving the industry to search for more inexpensive labor as evident from the large number of women employed in this sector (60% of 94 million total workers) who face a gender pay gap of up to 20% (International Labor Organization, 2023). However, this industry also plays a key role in the total earnings of middle- and low-income countries. Therefore, it is important that the challenges of environmental and social sustainability be addressed without eliminating these countries from the supply chain (World Trade Organization, 2022).

Further, the textile industry is responsible for 3–10% of global greenhouse gases, is one of the highest water consumers, and generates millions of tons of waste. Only 1% of textile waste undergoes fiber-to-fiber recycling to make new textile products, while the rest of it is downcycled for other applications (such as fillers, energy), disposed locally, or transported back to lower-income countries where it is landfilled without a potential for material recovery (McKinsey & Company, 2022).

The circular economy (CE) transition has emerged as an economic system to minimize waste generation, optimize resource efficiency, and promote sustainable production and consumption in this sector (Niinimäki, 2018). The CE framework is based on interrupting the linear model at various stages, as shown in Fig. 1. The

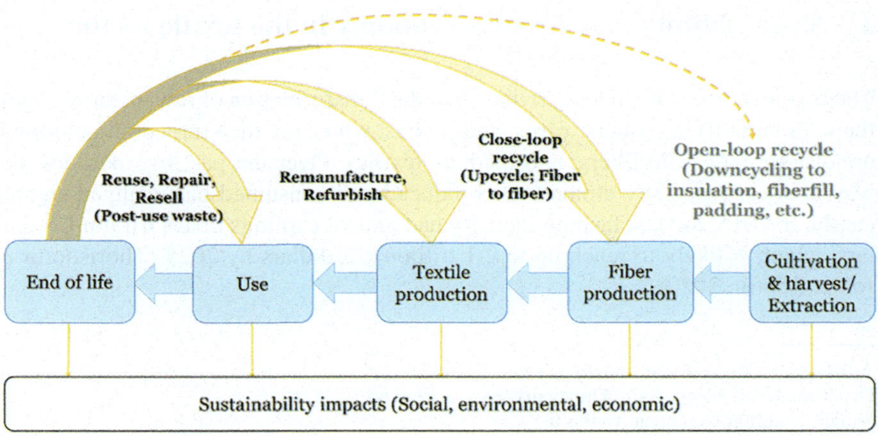

Fig. 1 Closing the loop in the textile and clothing industry—replacing the linear supply chain (in blue) with a circular framework (in yellow) to ensure material use efficiency

9R-framework is most used for understanding a transition to CE (Potting et al., 2017). The activities in this framework are R0 refuse, R1 rethink, R2 reduce, R3 reuse, R4 repair, R5 refurbish, R6 remanufacture, R7 repurpose, R8 recycle, and R9 recover. In the design stage, the CE framework proposes to refuse conventional production methods, rethink/redesign the supply chain, and reduce waste. In the consumption stage, users may reuse (or resell), repair, refurbish, remanufacture, and repurpose discarded products with or without technical intervention. At the end of life, the textile or garment may be recycled either completely or part of the material may be recovered to substitute virgin fiber.

Despite the interest in transitioning towards CE, a challenge for the textile industry to increase fiber-to-fiber recycling is the low efficacy of material separation and chemical recycling technologies (McKinsey & Company, 2022). This is further exacerbated by the challenge of collecting textile waste and the absence of design guidelines enabling easy garment recycling at the end of the product's lifetime (Niinimäki & Karell, 2020). On the consumer side, the uptake of circularity practices such as buying clothes made from recycled fiber or engaging with second-hand clothing markets is hindered by a lack of awareness regarding the safety of recycled content, associations of shame (cultural acceptance), or a lack of knowledge about the platforms (Mukendi et al., 2020).

The transition to CE is expected to reduce global waste generation, including textile waste, by 30%, generate local employment, and make goods accessible to new markets (Circle Economy, 2023). However, in this transition, issues of social sustainability have not been resolved (Ferioli et al., 2022). Many low-income countries face challenges like a lack of skilled workforce and skill development opportunities, putting pressure on factory owners to manage costs effectively. Some companies seem to only focus on avoiding child labor and sweatshops while foregoing other issues in working conditions in their suppliers' facilities. Also, many companies hire workers from subcontractors who do not have transparent information-sharing mechanisms with clear codes of conduct (Księżak, 2017). Companies also do not mandate their suppliers to abide by international standards (ISO 9001, ISO 14001, OHSAS 18001, SA8000) created to ensure ethical labor practices, and neither do they audit the suppliers adequately. However, the sustainability of an enterprise should be judged based on the sustainability of its supply chain.

Identifying and addressing the negative future impacts on stakeholders in the transition from a linear to a circular economy is also important since 9–12% of the global working population is employed in the apparel value chain (Suarez-Visbal et al., 2022). Further, countries like Haiti, Cambodia, and Bangladesh have more to lose with changes in business practices of textile production than the markets where goods are consumed. For example, in 2021, textiles were the most exported product from Haiti and significantly contributed to its industrial output, but it was only the 60th largest exporter of textiles globally (The Observatory of Economic Complexity, 2021).

Here, a textile company's corporate social responsibility (CSR) policies integrated with social life cycle assessment (S-LCA) may help identify potential benefits and hotspots of the CE transition. CSR entails companies' duty to create economic value while safeguarding society and the environment. CSR policies should surpass legal requirements and lead to societal advantages. Social issues targeted in CSR activities may include eradicating child labor, generating jobs, providing safe living conditions for communities, access to safe drinking water for workers, and appropriate training opportunities (Kumar et al., 2020). Here, S-LCA offers a structure for identifying these social issues over the complete life cycle of textiles, spanning from the extraction of raw materials to production, usage, disposal, and reintroduction into the supply chain through the CE framework.

With the rapidly growing textile market, various stakeholders like workers in the industry, communities settled near manufacturing sites, and consumers face risks related to unfair working conditions, toxic exposure, and lack of transparency in the supply chain. Hence, in the current chapter, the authors look at the role of CSR and S-LCA in promoting social sustainability. In Sect. 2, we will delve into CSR development strategies, how industries evaluate sustainability, and its implementation potential. In Sect. 3, we will discuss the methodological development of S-LCA and barriers to its application in this sector. In Sect. 4, we will discuss the potential to integrate S-LCA with CSR so that more targeted policies can be framed, and the role that government regulations play in promoting CSR. In Sect. 5, we present results from several cases where CSR policies could be based on social issues identified by the S-LCA, and in Sect. 6, we discuss the future of this integration.

2 Corporate Social Responsibility

The European Commission defined CSR in 2001 as "a concept whereby companies integrate social and environmental concerns in their business operations and in their interaction with their stakeholders on a voluntary basis." Therefore, CSR is a strategy for industries to "give back" to society, including people and the environment. Here, "people" refers to stakeholders like workers, customers/consumers, suppliers/retailers, and societies/communities. The term "people" also refers to shareholders, the government, investors, and competitors within the sector. Commonly, CSR policies are directed toward employees to ensure that they are provided with fair wages and decent working conditions without discrimination (Saxena, 2021). But these policies also provide a fair picture of a company's financial position to its shareholders. CSR activities may prove to the government that an organization will ensure legality in all its operations. It also shows customers that the organization is mindful of their product's impacts. CSR policies may assure suppliers of the ethical responsibilities such as fair competition and respecting their intellectual property rights. For overall society, a company may use its CSR policies to signal opportunities for progress through employment creation and philanthropic activities.

2.1 Development of CSR strategies

Developing CSR policies is not a systematic approach since it is a voluntary commitment of an organization that has the freedom to pick an issue they wish to address based on various drivers. The internal drivers may be a company's size, position in the value chain, and level of internationalization (Księżak, 2017). The latter point is important because, as mentioned in Sect. 1, large supply chains often have low levels of transparency potentially leading to more social risks. The external drivers for CSR may be environmental risks, regional/governmental push, the level of circularity expected, sustainability implementation challenges across the supply chain, and consumers' attitude toward sustainable clothing (Abbate et al., 2023).

In 2010, the ISO 26000 guidelines were published with an attempt to harmonize the key features of a CSR policy. Although it is not a mandatory standard, it encourages businesses to develop CSR policies that ensure transparency, respect human rights, meet stakeholder expectations, comply with laws, and promote sustainable development (Szewczyk, 2016). In 2013, the Indian government mandated companies to allocate 2% of their profits to CSR activities like poverty alleviation programs, women's empowerment initiatives, education for workers' children, and health checkups for employees (Chawla, 2019). Post the Rana Plaza collapse in 2013, Bangladesh's ready-made garment industry's (RMG) CSR initiatives have included a push to ensure workplace safety and compliance monitoring, social standards training for factories, sustainable use of natural resources, and improving worker benefits such as fair wages, skill development, gender equality, and better working conditions (Khattak et al., 2017). Efforts have also been made to promote collective bargaining rights for employees, leading to an increase in the reach of trade unions in the RMG industry from 138 in 2012 to around 800 in 2020. Other global CSR strategies include promoting consumer engagement by providing repair and resale options (Patagonia's "Worn Wear" and Eileen Fisher's "Renew" initiative), women empowerment through training opportunities for female workers (Gap Inc.'s "Personal Advancement & Career Enhancement" Program), ethical sourcing of raw materials like cotton (H&M's "Conscious Collection" campaign), spreading societal awareness about plastic pollution (Adidas's "Parley for the Oceans" program), and promoting better labor practices in the supply chain (Puma's "FOREVER BETTER" campaign).

2.2 Barriers to creating CSR strategies

Govindasamy and Suresh (2018) prioritized the motivations and barriers to CSR adoption (listed in Table 1) in the textile industry based on their interviews with 26 textile, knitting, and garment industries operating in Malaysia, out of which 18 were small and medium enterprises, and the remaining 8 were multinational companies. Based on the responses, they reported that the barriers to CSR adoption were costs,

Table 1 Highest to lowest priority of the motivations, barriers, and CSR initiatives based on interviews with 26 textile companies in Malaysia

CSR initiatives	Drivers	Barriers
Fair treatment for all employees and job applicants	Customer satisfaction	Lack of training for CSR knowledge
Reduce human rights abuses	Employee satisfaction	Insufficient financial resources
Enforcing ethical relations with suppliers	Sales maximization	Implementation too expensive
Enforcing ethical relations with shareholders	Protecting brand reputation	More important priorities for the company
Empowering women and other groups	Fulfilling government requirements	Implementation too complex
Support government policies	Market access	No significant benefits from implementation
Not harming the environment	Leadership in CSR	Difficulty in getting information on CSR implementation
Increasing community's economic stability	According to company's ethical principles	Lack of knowledge about CSR implementation
Supporting community development	Top management belief in CSR	Lack of knowledge about CSR practices
Solving social problems (problems were undefined)	Correct thing to do	Inadequate training from management for implementation
		Lack of worker-level skill/education for CSR implementation
		Implementation too time-consuming
		Lack of support from middle management
		Lack of support from top management

Modified from Govindasamy and Suresh (2018)

skills, awareness, management, size of the company, and lack of national/regional policies.

Similar observations were made in other parts of the world regarding CSR uptake. In discussions with Indian companies, a common response from CSR managers was that despite the mandatory CSR, companies are not always aware of which societal issues should be targeted (Saxena, 2021). Further, since CSR is mandatory, governments interfere or "direct" spending toward specific programs they wish to promote in a region (Saxena, 2021). Therefore, CSR activities may often remain a promotional exercise without significant positive effects on society.

The availability of funds is a barrier for CSR implementation in all industries. However, smaller companies are often reluctant to hire specialists since they can identify social challenges easily due to shorter supply chains and resolve issues through better communication with their suppliers (Walker et al., 2021). Nonrecurring costs of implementing CSR are related to the infrastructure needed to produce sustainably and donations for philanthropic activities. Recurring costs include regular payments (or remunerations) for social activities and periodic certifications and trainings for inexperienced staff members. There are also costs related to the designations for people who will create and manage CSR activities such as directors and managers. Finally, there are expenses related to hiring responsible contractors and subcontractors, cooperating with local governments, enforcing government regulations, and investing in the mitigation of environmental damage.

The CSR managers in the Netherlands and Italy suggest that worker rights' violations should be within the ambit of the supplier country's own laws and not the responsibility of an individual business, since not all countries abide by the same international labor practices (Walker et al., 2021). For instance, in cases where overtime is accompanied by extra wages, implementing the SA8000 rule of limiting overtime to 12 h will restrict hourly wages, and workers may find that unacceptable, even though excess fatigue could reduce productivity in the longer term (Księżak, 2017). Also, the scope of altering worker conditions, brand image, and green supply chain management initiatives depends on the company's own interests, though national interests such as anti-sweatshop movements or extended producer responsibility may push the adoption of voluntary codes of conduct in the fashion industry and influence CSR policies (Thorisdottir & Johannsdottir, 2020).

The introduction of CSR rules in the textile and apparel industry is also hindered by the lack of control caused by geographical distance between producers and users, which may lead to low levels of awareness in focal companies about the social hotspots in the supply chain. For instance, the Nike sweatshop scandal of the 1990s and 2000s highlighted the disconnect between the brand and the workers involved in the manufacturing of their products (Burhan Wazir, 2001). A survey of audit results shows that change in supplier behavior depends on the ease of measuring an impact category. For instance, child labor or health and safety (number of accidents) are much easier to spot in the supply chain than whether employment was freely chosen or forced (Meier, 2015). Forced labor is more difficult to identify because of inconsistencies in the collected data for wages or working hours (Shaikh et al., 2020). Inconsistencies include incorrectly recording time of work like no entrance/exit time being recorded, or time stamps being too early/late, or being intentionally falsified. Despite the push toward using audits, falsification of data is a common challenge faced by apparel companies (Judd & Kuruvilla, 2020). Hence, companies need to collaborate closely with suppliers to support their capacity-building efforts and promote responsible practices. Suppliers and subcontractors should be encouraged to obtain recognized certifications and be open to audits that clearly demonstrate their credibility, which entitles them to recognition as a responsible and sustainable participant in the sector. Audits must be accompanied by pressure to publish reports as per the ILO conventions followed by a dialogue with stakeholder groups regarding the identified issues.

Further, commitment toward CSR depends on the perceived exclusivity (Chan et al., 2020). Hence, Chan et al. (2020) suggest that it is better for fast fashion brands (e.g., H&M) and brands for functional garments (e.g., Nike) to commit a higher CSR fund compared to a luxury fashion brand (e.g., Louis Vuitton). The functional brands should seek improvements in product safety, such as using nontoxic dyes, and maintain a high CSR commitment level to ensure a positive brand image. For the fast fashion brand, it is recommended that they set a premium price for clothes made from sustainable raw materials like organic cotton, such as their "conscious" line. However, luxury brands are often not associated with sustainability and hence should have a higher advertisement level while adopting measures to report their ethical sourcing practices more transparently.

Some researchers (Byrd & Su, 2021; Chawla, 2019; Su et al., 2019) have also reported that when adopting CSR policies comes at a price for consumers, there is hesitancy in clothing companies to transition due to fear of loss in sales. Although companies that are more consumer-facing tend to promote social sustainability to improve their reputation and gain a competitive advantage. For instance, clothing brands have stricter CSR policies than yarn and textile manufacturers, and larger firms (by number of employees) often have more comprehensive CSR practices than smaller ones due to higher public scrutiny (de Abreu, 2015). However, identifying CSR as a form of public relations exercise can lead firms to engage only in actions that improve the image of the firm (Szewczyk, 2016).

2.3 *Evaluating the social sustainability in CSR*

An organization's commitment to CSR initiatives in the textile and clothing industry can often be quantified using sustainability indices. Several indices are available for evaluating material recovery and circularity like Cradle to Cradle® (McDonough & Braungart, 2002), the Business Social Compliance Initiative (BSCI, 2013) and environmental lifecycle assessment (ISO 14040, 14044). However, only a limited number of indices discuss social sustainability, such as:

1. *Higg Facility Social & Labor Module (FSLM):* The Higg FSLM advocates for enhancing fair labor conditions globally within a value chain. Rather than focusing solely on compliance, businesses could pursue systematic changes that positively influence the well-being of individuals across the entire supply chain. Leveraging the unified framework of the Social & Labor Convergence Program (SLCP), the Higg FSLM is utilized by entities such as brands, retailers, and manufacturers to evaluate the working conditions of those contributing to the production of numerous consumer goods annually. The FSLM assesses components like management systems, operational controls, oversight, and engagement with workers. The scoring mechanism is universally applicable, ensuring improved social performance in manufacturing facilities globally. Points are allocated based on universal queries, delivering prompt insights into potential risk areas and driving changes in actions, decisions, and practices.
2. *Global Reporting Initiative (GRI):* GRI provides a comprehensive framework for sustainability reporting, including reporting guidelines specific to the textile and clothing industry. The GRI Standards cover a wide range of environmental,

social, and governance (ESG) aspects, allowing companies to disclose their sustainability performance and impacts.
3. *Dow Jones Sustainability Index (DJSI):* DJSI is a globally recognized sustainability index that evaluates the sustainability performance of companies across various industries, including textiles and apparel. It also assesses companies based on environmental, social, and governance criteria, providing an indication of their sustainability, leadership, and performance.

Several indices are also used by certification organizations to measure sustainability. For instance, the Textile Exchange provides certifications and assessment frameworks based on the Organic Content Standard, Responsible Wool Standard, and Global Recycled Standard to ensure organic materials, recycled content, and animal welfare. The Carbon Disclosure Project also helps companies to measure and disclose their environmental impacts, greenhouse gas emissions, and other climate-related risks. However, a challenge with all these indices is the collection of accurate and comprehensive data, such as sourcing practices by suppliers or greenhouse gas emissions. This data collection may be more challenging for small and medium enterprises (SMEs) due to the additional time and costs involved. Further, larger brands may be reluctant to report data about their supply chains (Capasso, 2021).

Social life cycle assessment is a tool that has emerged out of the need to identify the risks to various people involved in the lifecycle of a product's development, use, and disposal. S-LCA may assist companies in conducting their operations in a socially responsible manner while acknowledging their need to remain competitive and profitable in the market (Dreyer et al., 2006). The identification of impact categories and relevant indicators is guided by universally accepted documents on social issues (like ILO, Human Rights Declaration) and the specific business context of each company as discussed in the next section.

3 Social Life Cycle Assessment

S-LCA is a tool to assess the impacts of a product or process on people such as workers, consumers, or society throughout its life cycle. The first official guidelines for performing an S-LCA were created by the United Nations Environment Program (UNEP) in conjunction with the Society of Environmental Toxicology and Chemistry (SETAC) in 2009. These UNEP/SETAC guidelines have been subsequently updated in 2013 and 2020. A detailed methodological sheet was published in 2021 to further explain the process of data gathering for an assessment (UNEP, 2021), and a pilot project report to show the successful implementation of these guidelines was published in 2022 (Traverso et al., 2022).

3.1 Development of S-LCA

S-LCA is based on the same framework that is used for the environmental life cycle assessment, i.e., the ISO 14040 and 14044 (ISO, 2006a, b) standards but has been modified to account for more qualitative social impacts using methods from social sciences (Grubert, 2018). Like CSR policy framing, the social impacts are evaluated among several stakeholders. In the UNEP/SETAC guidelines, these stakeholders are divided into six categories: workers, consumers, value chain actors, local community, society, and children. The positive and negative impacts are then classified by stakeholder categories as shown in Table 2. The most frequently used impact categories are related to workers, such as their health and safety, working conditions, child labor, and forced labor. The limits for "acceptable" social impacts are defined by agreements of the international community. For instance, the International Labor Organization defines the number of working hours acceptable without the labor practices being considered exploitative.

Figure 2 shows the steps involved in an S-LCA. It starts with defining the goal and scope, which includes setting the functional unit, system boundaries, and impact categories to be considered. The second step is to create a data inventory, which for textiles includes information on raw materials used, manufacturing processes, distribution strategies, use (and reuse), and disposal (with or without material recovery and recycle). All these lifecycle stages could occur in different geographic locations with several stakeholders being impacted simultaneously (Roos et al., 2016). Data sources can include company records, interviews with stakeholders, public reports, databases, and social audits. The third step is to assess the social impacts using established impact assessment methods, which involves categorizing the collected inventory against relevant impact categories (Table 2), such as worker rights, occupational health and safety, fair wages, human rights, community well-being, and stakeholder engagement. The fourth step is to interpret the results of the impact assessment and analyze the significance of the identified social impacts. This step involves comparing the findings against relevant benchmarks, guidelines, or standards to determine the extent of compliance or improvement needed. These results are reported transparently, including the identified social hotspots, areas of improvement, and potential solutions. The report should also include a discussion of the limitations and uncertainties associated with the assessment.

There are two approaches to S-LCA: the reference scale approach (RSA) and the impact pathway approach (IPA). The RSA helps determine social performance or social risk based on specific reference points of expected activity (called performance reference points—PRPs). These PRPs are based on data, information, or judgment provided in international agreements (ILO, UNHRC). It is helpful for evaluating the "potential social impact" but cannot be used to identify a causal link between the impact and an activity in the supply chain. Whereas the impact pathway approach is used for a quantitative analysis of the social consequences of a product system or process. IPA uses correlation/regression-based relationships between the activities of a supply chain and the resulting potential social impacts. This is like the

Table 2 Impact categories for each stakeholder given in UNEP/SETAC Guidelines

Workers	Local community	Value chain actors (not including consumers)	Consumer	Society	Children
Freedom of association and collective bargaining	Access to material resources	Fair competition	Health and safety	Public commitments to sustainability issues	Education provided in the local community
Child labor	Access to immaterial resources	Promoting social responsibility	Feedback mechanism	Contribution to economic development	Health issues for children as consumers
Fair salary	Delocalization and migration	Supplier relationships	Consumer privacy	Prevention and mitigation of armed conflicts	Children concerns regarding marketing practices
Working hours	Cultural heritage	Respect for intellectual property rights	Transparency	Technology development	
Forced labor	Safe and healthy living conditions	Wealth distribution	End-of-life responsibility	Corruption	
Equal opportunities/ discrimination	Respect for indigenous rights			Ethical treatment of animals	
Health and safety	Community engagement			Poverty alleviation	
Social benefits/ social security	Local employment				
Employment relationship	Secure living conditions				
Sexual harassment					
Smallholders including farmer					

"characterization" step in ISO 14040/14044 for environmental LCA, where process outputs such as CO_2 emissions are linked with resulting environmental problems like global warming and its consequent impact on human health. Hence, the analysis may be used to identify the consequences of activities leading to longer-term impacts. Due to its complexity, IPA is less frequently used than RSA for S-LCA (UNEP, 2020).

For a textile supply chain, S-LCA enables the assessment of factors like working conditions, human rights, and community well-being throughout the supply chain,

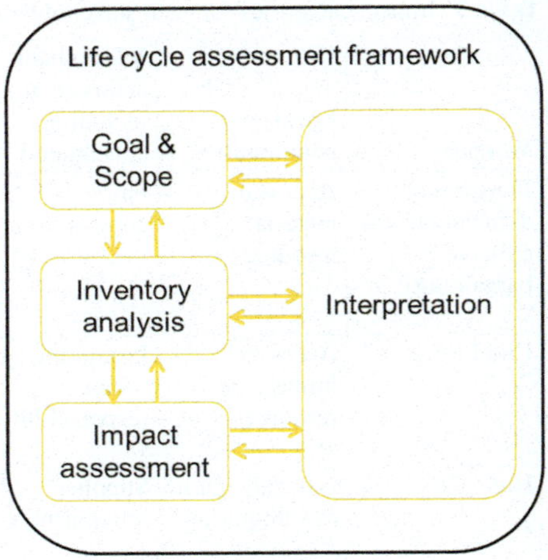

Fig. 2 Steps for performing a social life cycle assessment (From ISO, 2006a).

promoting ethical sourcing, improving labor conditions, and engaging in responsible disposal practices (Lenzo et al., 2017). S-LCA can identify the social impacts associated with circular practices (reduce, reuse, recycle), such as the employment opportunities created through recycling initiatives or the potential social challenges of transitioning to new business models (Yıldız-Geyhan et al., 2017).

Data requirements for conducting an S-LCA in the textile/clothing industry include information about the company's operations, supply chain structure, and practices related to working conditions, human rights, and community engagement. This can include data on labor policies, social initiatives, employee well-being programs, and community involvement. Data on suppliers and subcontractors involved in the production process, including their labor practices, certifications, and compliance with relevant standards or codes of conduct. Information about the stakeholders affected by the textile/clothing industry, such as demographics, socio-economic conditions, and potential impacts on their well-being. Reports from social audits conducted by independent third-party organizations provide insights into working conditions, compliance with labor rights, health and safety practices, and other social aspects within the supply chain. Data on industry-wide standards, regulations, and guidelines related to working conditions, human rights, and community engagement, regional labor laws, social policies, and cultural aspects can be relevant.

While sustainability indices like GRI, DJSI, and the Higg Index consider social aspects, the integration of S-LCA principles can enhance the assessment of social impacts in the textile industry. S-LCA provides a structured and systematic approach to evaluating social dimensions, enabling companies to identify social hotspots, improve labor practices, and promote stakeholder engagement more effectively within the context of these sustainability indices. S-LCA can serve as a valuable tool to fulfill the guiding principles of ISO 26000 by providing a systematic and comprehensive approach to addressing social responsibility within organizations.

3.2 Barriers to using S-LCA

There are several barriers to using the S-LCA tool for the textile sector. The first barrier is that of data collection. Since textile supply chains are spread widely, most assessments rely on secondary data collected through national statistics and reported in some generic databases. But the generic assessments have different scopes from those required for a particular site (Ekener-Petersen & Moberg, 2013), which makes the assessment unreliable. Another issue is the transparency of reported information in the databases. There is also a lack of agreements between all stakeholders on the prioritization of impact categories, low applicability of generic data to SMEs, and poor representation of low- and middle-income countries in generic databases (Fontes et al., 2018). The quality of the stakeholder responses should be checked using a scoring system, which calculates average scores from individual stakeholders by corroborating the response with available reports (Zerbino et al., 2023). In general, the global databases for secondary data collection (like PSILCA or SHDB) need to be continuously expanded with more localized case studies.

The second barrier is the inclusion of relevant stakeholders. While information related to workers can be captured through precise indicators, information related to communities, value chain actors, and consumers is difficult, and the indicators are vague. S-LCA also cannot capture the product's "use phase," despite its significance in extending product lifetime through options like repairing or reusing (Suarez-Visbal et al., 2022). S-LCA practitioners have used indicators like comfort, convenience, ease of use, and social acceptance to evaluate the "use phase" impact.

The third barrier is that the UNEP/SETAC guidelines may be differently interpreted based on each situation (Chhipi-Shrestha et al., 2015). For example, the CE transition may alter job skills due to changes in consumption patterns and production processes. Hence, there is likely a loss of jobs in resource extraction due to increased resource use efficiency, job substitution where some activities will be replaced by others, e.g., from landfilling and waste incineration to recycling, both of which require reskilling, which is not captured in the current assessment method (Borms et al., 2023). In addition, it is still not clear how to assess the potential social impact of materials with multiple lifecycles in terms of impact attribution across the various life cycles (Papo & Corona, 2022). Standard indicators should be created based on the global consensus of socially relevant themes such as the Sustainable Development Goals (Herrera Almanza & Corona, 2020) or through collaboration with businesses regarding context-specific requirements (Walker et al., 2021).

Based on Sects. 2.2 and 3.2, there are several challenges in developing CSR and S-LCA. CSR activities may be focused on any number of issues from environmental, societal, to economic. A company that is setting aside funds for CSR activities wants to include activities that either directly benefit it or at least create a perception amongst stakeholders of the companies' ethical leanings. It is used as a signal rather than a problem-solving tactic which leads to limited application of the funds for influencing society, and other stakeholders. CSR managers are also unaware of aspects concerning employees' welfare or improving working conditions in the firm. The biggest challenge in S-LCA is often associated with data collection, verification, and standardization. However, both these tools can be improved through collaboration and engagement among different stakeholders, including suppliers, customers, industry associations, NGOs, and local communities. Expert interviews, surveys, and questionnaires are often used in this step to ensure open dialogue that promotes knowledge sharing regarding social issues effectively (Aparcana & Salhofer, 2013; Staicu & Pop, 2018). This step would be further enhanced through integrating S-LCA into CSR framework.

4 Integrating S-LCA in CSR

The textile industry has focused on promoting sustainable business models through supply chain innovations, which includes promoting the use of environmentally friendly materials and ensuring that suppliers follow a code of conduct that ensures workers' safety. However, it has been discussed in previous sections that there is a need to broaden this focus and include all stakeholders across the lifecycle of textiles, which could be done through an S-LCA.

An effective assessment involves defining what is to be safeguarded under regulatory requirements for a region, the boundary of the supply chain based on the function of the product under evaluation, identifying the role of all stakeholders involved in the lifecycle of the product, and collecting product and site-specific information wherever possible. The results should then be used in creating a CSR policy that ensures negative impacts are minimized and positive impacts are maximized. For instance, S-LCA can identify hotspots in the supply chain such as child labor and by leveraging CSR initiatives to direct funds toward establishing schools and opportunities for the workers' children to be engaged safely while getting benefits of being close to their parents/guardians. S-LCA can also guide decisions on material selection, manufacturing processes, supply chain optimization, and product design for improved social sustainability.

Additionally, since CSR reporting is not always verifiable or standardized, S-LCA can help answer questions like how far the supply chain responsibility goes and where to find the solutions. For instance, when evidence emerges of unfair labor practices through S-LCA, companies can make the effort to identify better suppliers who follow the expected code of conduct and make the audit reports transparently available. Hence, S-LCA promotes transparency and accountability by engaging stakeholders and considering legal and regulatory compliance to ensure all business practices align with global expectations. The remaining section will highlight factors that could help integrate S-LCA with CSR, such as regulations, managers, employees, and consumers.

4.1 What is to be safeguarded: Defining areas of protection through regulations

International, national, and regional regulatory efforts play crucial roles in promoting social responsibility within the textile sector. International bodies like the United Nations, ILO, and the Organization for Economic Co-operation and Development (OECD) set global standards and guidelines for social responsibility, with the ILO's Fundamental Conventions establishing core labor rights. Initiatives like the OECD Due Diligence Guidance for Responsible Supply Chains encourage transparency and responsible sourcing, guiding companies in managing social risks across complex supply chains.

At a national level, many countries enact laws addressing labor rights, workplace safety, and related social issues in the textile industry (de Abreu, 2015). For instance, Pakistan's Protection against Harassment of Women at the Workplace Act, 2010 provides a procedure to file a complaint for sexual harassment in the workplace. The Equality Act of the United Kingdom, enforced in 2010, protects employees against discrimination in the workplace at every stage of employment. The Fair Labor Standards Act of 1938, passed in the United States, gives a provision for minimum wage and overtime pay for work over 40 h. Additionally, regulations governing the import and export of textiles often incorporate social responsibility criteria, discouraging practices like forced or child labor. Governments also establish labor standards and conduct audits to verify compliance with international and national labor laws. Environmental regulations such as the Clean Water Act (US, 1972) and the National Air Quality Action Plan (China, 2014), though primarily focused on the environment, indirectly promote social responsibility by impacting worker health and community well-being. Tax incentives and other benefits may also be offered to companies demonstrating strong social responsibility practices. Further, public procurement policies in some countries favor suppliers with verifiable social responsibility records, encouraging responsible practices to access public sector markets. Examples include the Act on Public Procurement and Concession Contracts implemented in Finland in 2016 and the Public Procurement Act from the Netherlands implemented in 2012. Industry associations and coalitions work alongside regulatory bodies to develop and enforce social responsibility standards within the textile sector. Therefore, regulatory efforts provide a comprehensive framework for advancing social responsibility in the textile industry, establishing clear expectations, incentivizing responsible practices, and holding companies accountable for their social impacts. This contributes to a more sustainable and ethical industry overall.

Advancing CSR that is integrated with S-LCA requires establishing clear standards and expectations for ethical social and environmental conduct in business practices. Several regulatory efforts have been undertaken, which determine the compliance strategies that businesses have to adhere to while framing CSR policies, and these also aid in creating the PRPs for RS-S-LCA. In this section, some of the policies are listed.

4.1.1 Human Rights Due Diligence

The UN Human Rights Council's Guiding Principles on Business and Human Rights, published in 2011, promote Human Rights Due Diligence (HRDD), which assigns clear responsibilities to governments and businesses to protect human rights. HRDD is used by companies to frame CSR policies that ensure human rights in all their global activities, relationships, and value chains. The HRDD, coupled with robust corporate endorsement of the Sustainable Development Goals, has helped companies to identify, prioritize, and address social risks within their value chains. HRDD has also played a significant role in enhancing social responsibility by helping companies demonstrate their commitment to meeting legal and ethical obligations while also respecting and safeguarding human rights across their operations, supply chains, and interactions. HRDD encourages organizations to engage with a broader range of stakeholders, including affected communities, workers, NGOs, and local authorities, to identify and address potential human rights risks ensuring transparency and stakeholder inclusion in CSR. It strengthens risk management processes and ensures a more comprehensive approach to framing CSR initiatives. HRDD broadens the focus on the entire value chain, from suppliers to customers, in assessing human rights impacts and underscores the importance of responsible sourcing and production. HRDD also requires organizations to disclose information about their efforts to identify, prevent, and address human rights violations, which enhances accountability and promotes ethical practices. Incorporating HRDD into CSR policies helps companies maintain a positive brand image and reputation by demonstrating their commitment to respecting human rights, which is of increasing importance to consumers and investors. HRDD aligns with the SDGs, particularly Goal 8 (Decent Work and Economic Growth) and Goal 16 (Peace, Justice, and Strong Institutions), fostering constructive collaboration between CSR efforts and global sustainable development priorities.

4.1.2 Circular Economy Action Plan in the European Union

In the European Union, the Circular Economy Action Plan (CEAP) was introduced in 2020. It primarily focuses on reducing environmental footprints through improved raw material consumption, promoting recycling and upcycling, minimizing waste generation throughout the product life cycle, using eco-friendly materials, and addressing issues like chemical management, water consumption, and greenhouse gas emissions. These measures will target the health and safety of various stakeholders. However, it also includes increased transparency and traceability measures that will help address issues like unethical labor practices. CSR policies can play a crucial role in the adoption and implementation of the CEAP and the EU Strategy for Sustainable and Circular Textiles by driving internal commitment toward sustainability and circularity within the textile supply chain.

4.1.3 Extended Producer Responsibility

The fashion industry's transition to an effective and sustainable circular economy requires a complex mix of policy initiatives. Key to this is implementing an Extended Producer Responsibility (EPR) regime, shifting environmental impact costs from taxpayers to producers. The EPR schemes, already in place for packaging and electrical goods, are undergoing reforms, with new rules coming in 2024. The EU aims to extend EPR to fashion and textiles, seeking feedback from industry stakeholders through consultation (ongoing work in years 2022–2024). While the industry supports EPR, the cost should incentivize sustainable practices, penalizing harmful or difficult-to-recycle materials. Eco-modulation can encourage eco-friendly choices, discouraging excessive waste. Eco-labeling can inform consumers about garment durability, repairability, and recyclability. Revenue generated by EPR could be reinvested in the industry to support eco-design principles and improve recycling infrastructure. By banning textiles from landfills and moreover promoting recycling and innovative practices, we can move toward a more sustainable balance in the textile and fashion sector.

4.1.4 EU Sustainable Textile Strategy

The EU, European regulators have had a strong motivation to push industrial development toward CE, and lately, the focus has been on the textile sector. Especially the huge growth in environmental impacts caused by imported fast fashion has caused a discussion. The European Commission has initiated the *EU Strategy for Sustainable and Circular Textiles* along with CEAP (Directorate-General for Environment, 2022).

The EU Strategy for Sustainable and Circular Textiles promotes steps that may reduce the negative impact of textile products over their entire life cycle and support green and digital transitions in the sector. The EU Strategy will also boost textile waste collection and recycling. It will also provide a "mandatory minimum" regarding the quantity of recycled fibers to be used in all textile products and encourage companies to make products with longer lifetimes by making them easier to repair and recycle at the end of life. Apart from these, the policy is targeting many environmental initiatives such as preventing the release of microplastics and destroying unsold/unused garments. It also provides tools to control "greenwashing" to protect consumers and other environmental and labor concerns across the value chain. Policy goals can increase local employment (e.g., in garment repair business, in textile collection and sorting), yet there is a risk that some of these works will move to developing countries, especially the sorting and recycling of textiles. Currently, "the Global South produces fibres and textiles, and garment assembly work is located there, but the garments are produced for consumers in the Global North. The notion of economic development is rooted in Western colonial capitalist narrative and this grounding prevents improvements to the social side of sustainability" (Niinimäki, 2023). Accordingly, there is a risk that the Western world transports the social impacts of CE and textile recycling to other countries, continuing the colonial message and power play run by the Western world. Used clothing from the EU is exported to Ghana, Ukraine, Pakistan, the United Arab Emirates, and Nigeria with the aim that these garments can be used again, yet the reality is that only 30–65% of used clothing can be resold, and the rest is dumped or burnt (EEA European Environment Agency, 2023). Targets such as improving gender equality and improving worker conditions fall under the ambit of CSR initiatives that could be undertaken by a company.

4.1.5 Corporate Sustainability Due Diligence Directive

The Corporate Sustainability Due Diligence Directive (CSDDD) was introduced in 2023. It directs large companies to reduce the negative impacts on human rights and the environment occurring in their own operations and the associated supply chain. The specific initiatives under CSDDD could be related to labor upskilling or reskilling and training for new skills, which improve diversity in the labor.

In the past decade, there has been an attempt by organizations to direct CSR toward fulfilling the sustainable development goals (Kumari & Singh, 2023). The three SDGs that may be targeted by textile companies through their CSR efforts are sustainable production and consumption (SDG12), reducing water consumption (SDG6), and ensuring decent working conditions (SDG8).

4.2 Who Is to Be Safeguarded: Defining Stakeholders

The primary difference in the strategies for framing CSR policy vs an S-LCA method is the definition of stakeholders. While CSR defines stakeholders as groups or individuals that can affect or are affected by the activities of an organization (de Abreu, 2015), the UNEP/SETAC guidelines suggest that stakeholders include anyone who can influence or is influenced by the activities taking place over the entire lifecycle of a product or process (UNEP, 2020). The role of stakeholders in framing CSR policies is listed in this section.

4.2.1 Role of Managers

Managers often prioritize CSR policies for optimal allocation of resources. Most of the initiatives for CSR have focused on environmental initiatives, including waste management, reducing pollution, minimizing chemical discharge, organizing tree plantation drives, and other "cleanliness drives" such as for rivers, beaches, or public parks. Other than this, CSR policied have focused on better material procurement through organic seeds purchased for cotton cultivation. Companies are also striving to ensure no discrimination based on race, religion, gender, and other perceived forms of identity. Several managers responded that although child labor exists, it is a result of poverty rather than an intent of hiring young workers (<12 years) and is actively being reduced by companies (Chawla, 2019).

4.2.2 Role of Consumers, Local Communities, and Society

Consumerism has been a key driver in the fashion industry that is often related to the "aspirations of symbolic lifestyle," which makes it difficult change consumer practices (Thorisdottir & Johannsdottir, 2020). However, a closer look at consumer attitudes as a driver for sustainable fashion practices shows that there is a concern about the supply chain transparency, especially regarding the ethics of material sourcing and the social effects on manufacturing workers and communities (Byrd & Su, 2021). In terms of age-based differences, the younger generation chooses affordability before the environmental and social concerns, but there is a positive attitude across all ages and nationalities toward ensuring sustainability in the clothing industry (Byrd & Su, 2021). In another study, Su et al. (2019) report that millennials (born between 1980 and 2000) from the United States are more aware of sustainability issues in the apparel industry than their Chinese counterparts. Therefore, brands could focus on educating all consumers about their sustainability practices to ensure consumer loyalty. Regional differences were evaluated by Chawla (2019), who reported that European consumers (specifically, French, German, Slovene, and Swedish) were more likely to support responsible businesses compared to their US counterparts. Indian consumers also expect fair treatment of workers/employees but are more likely to value philanthropic activities than legal or ethical CSR activities. Consumers interested in sustainability are willing to pay a premium for goods made from socially responsible practices.

4.2.3 Role of Workers/Employees

Although textile industries focus on clients and consumers, they should also ensure employees and other stakeholders (shareholders and suppliers) are prioritized since CSR policies that focus on treating employees fairly with provision of funds for education, social welfare, healthcare, skill development, and women empowerment create a positive attitude in employees and in their decision to work in an organization. However, in the textile sector, farming communities have been ignored by CSR policies, leading to poverty, lack of clean drinking water, low standards of living, and unemployment (Chawla, 2019).

5 Case Studies for S-LCA in Textile Industry

Several cases have reported the successful application of S-LCA for identifying social hotspots in textile supply chains. The key learnings from these studies are presented in this section and Table 3, based on which companies may identify their social responsibility.

In the case of SMEs, which are the primary employers in a region, there is a social responsibility to safeguard their workers and communities (Lenzo et al., 2017). But there is an issue of reliable data for suppliers and consumers, especially at the small business level. However, the lengthy textile supply chain means no company can prevent all the social problems at a national or global scale with its secondary suppliers and business partners.

When comparing handloom industries with power loom (SMEs and large industries), it was reported (Mahiat et al., 2022) that handloom was more sustainable both environmentally and socially. The handloom industries were more socially inclusive than the power loom industries, especially regarding employment opportunities. Specifically, handlooms offered more local employment, heritage conservation, and female employment, but the rate of migration associated with handlooms was higher, as there were only a few of these operational in Bangladesh (at the time of publication) since power looms offered a larger scale of production. Conversely, power looms offered better safety and working conditions, including a fixed working duration.

Muñoz-Torres et al. (2023) reported that in the textile sector, workers' safety is typically a bigger concern in the supply chain than other issues like collective bargaining and fair income. This was also reported by Van der Velden & Vogtländer (2017). Corruption was also highlighted as a major issue in the supply chain, especially in the phases taking place in low-income countries and in parts of Eastern Europe (Martin & Herlaar, 2021). Zamani et al. (2018) used the Social Hotspot Database to evaluate the textile supply chain and reported that persistent sectoral-level issues for workers include child labor and fatal and nonfatal injuries due to low operational safety, and for local communities, it is the likely exposure to toxins or hazardous chemicals. Herrera Almanza & Corona (2020) reported that social risks were more concentrated in the manufacturing sector, especially during the manufacturing of shirts in Bangladesh, fabric in Myanmar, and spinning activities in China. These risks were either environmental or related to trade unionism, low levels of social responsibility, long working hours, poor living conditions, low salaries, and reduced access to immaterial resources for the community. An observable difference in worker safety was reported when the supply chain of a Swedish sweater manufacturer moved its production partly to Europe by incorporating circular practices, such as using waste wool generated in sheep farms to replace virgin wool fibers obtained from Asia. All phases of production showed an improvement in social sustainability (Martin & Herlaar, 2021).

A common theme that emerged from all the S-LCA cases is that low-income countries have unfair labor practices. Typically, low wages, long working hours, and

Table 3 Published case studies on application of S-LCA to identify social responsibilities for textile industries

Source	Stakeholder evaluated	Location of study	Size of company	Functional unit/ reference unit	Key results	Potential CSR initiatives
Lenzo et al. (2017)	Workers and local community	Sicily, Italy	SME (local business)	A garment consisting of 495 capes, knitted in a soft blend of wool and cashmere (60% wool and 40% cashmere)	Local employment generated; community had a positive attitude toward the company; challenges in obtaining reliable data for suppliers and consumers	Enforcing ethical practices by suppliers such as fair wages and decent working conditions
Mahiat et al. (2022)	Society, workers, local community, and value chain actors	Bangladesh	Handloom	1 ton of 100% cotton-woven fabric	Handloom industry was more inclusive than power loom industry and offered more local employment, heritage conservation effort, and female employment. Power looms performed better on rate of migration (i.e., lower migration); more safety and better working conditions	Obtain certifications to show environmental sustainability, fair labor practices, and cultural importance (through GI tagging). Handloom skill training for communities. Women empowerment programs
Muñoz-Torres et al. (2023)	Labor rights, health & safety, human rights, governance, and community	Global	Unspecified	1 cotton t-shirt (unspecified size/ weight)	Worker's safety, collective bargaining, wage assessment; corruption during manufacturing in low-income countries	Reduce human rights abuses and direct funds toward sustainable sourcing
Van der Velden and Vogtländer (2017)	Workers	Global	Unspecified	Unspecified	Low wages in Bangladesh and Myanmar during production	Enforcing ethical practices by suppliers such as fair wages and decent working conditions

(continued)

Table 3 (continued)

Source	Stakeholder evaluated	Location of study	Size of company	Functional unit/ reference unit	Key results	Potential CSR initiatives
Zamani et al. (2018)	Labor rights, health & safety, human rights, governance, and community	Sweden	Unspecified	1 USD worth of clothing	Child labor, injuries, and fatalities (operational safety), and toxics and hazards (for local community)	Promote community's economic stability, philanthropic activities in supplier countries like schools and skill development institutes
Herrera Almanza & Corona (2020)	Workers, local communities, value chain actors, and society	Netherlands	Multinational company	A white, long-sleeve, medium size (M) man's shirt, of 243 g weight, made of 97% cotton and 3% elastane, worn for working or special events (low-frequency usage)	Low rate of trade unionism, low levels of social responsibility, long working hours, poor living conditions, low salaries, and reduced access to immaterial resources for the community	Enforcing ethical practices by suppliers such as fair wages and decent working conditions through audits and checks
Martin and Herlaar (2021)	Workers	Sweden	Multinational company (Röyk)	One mid-weight sweater (600 g)	By partly shifting the supply chains from Asia to Europe, Röyk reduced their environmental footprint through lower transport needs, avoided forced labor, but risk of corruption observed in Eastern Europe in wool processing phase	Sustainable material sourcing and certifications for use of recycled product safety
Roos et al. (2016)	Workers	Sweden	Unspecified	1 USD worth of apparel for the Swedish market	Unfair wage, child labor, and exposure to carcinogens at workplaces in the production phase	Roadmap for suppliers to improve workers' living conditions, paying a fair living wage; implement pay structures at all factories; engage with governments on minimum wage issues; and update the company's code of conduct

poor safety equipment were reported. Often, "factory workers and their families are malnourished, cannot afford proper education for their children or proper medical care, have poor living conditions and no access to clean drinking water" (Niinimäki, 2023, p. 86).

Table 3 lists some of the potential CSR initiatives that companies can take to address the social risks identified in their supply chains. Typically, companies that have identified issues of worker safety or forced and child labor in their supply chains should either identify new suppliers or at least enforce a strong supplier code of conduct. Additionally, CSR efforts should go into philanthropic activities such as setting up schools for workers' children to save them from falling into practices like child labor. CSR policies should focus on engaging suppliers who provide better weekly hours, overtime hours, freedom of association, gender equality, and waste management systems.

The companies operating in global supply chains for apparel manufacture must be more concerned about material sourcing especially for cotton and should frame CSR activities that direct funds toward sustainable sourcing. Companies like Röyk, Sweden (Martin & Herlaar, 2021) could encourage other businesses in their region to identify risks/improvements in their social sustainability through S-LCA. Handloom industries should focus their CSR funds on obtaining certifications that show their environmental sustainability, fair labor practices, and cultural importance through "geographical indication" or GI tagging. Additionally, they can organize skill development at community level, create women empowerment programs, and advertise about the exclusivity and sustainability of the product (Mahiat et al., 2022).

Only one example showed a successful use of S-LCA for framing CSR policies. Results from Roos et al. (2016) were used by Swedish brand H&M to create their Fair Living Wage roadmap for improving workers' living conditions. Specifically, they provided support to suppliers for paying a fair living wage, organized pilot projects for testing out a fair wage method in their factories in Cambodia and Bangladesh, implemented new pay structures, involved state organizations in setting the minimum wage, and updated their code of conduct.

Based on the S-LCA results, CSR policies could focus on various driving forces, which could be categorized into internal, natural environment, relational, and discretionary. Internal practices focus on providing a healthy and safe environment and appropriate career plans for employees, guaranteeing the production of a quality product. The social impact categories related to this would be the ones listed under "worker" stakeholder in Table 2. The natural environment practices encompass an environmental management system identified by an environmental LCA. Discretionary practices include the development of social programs focused on local communities and may be evaluated using the categories listed under "local community," "society," and "children" stakeholders. Finally, relational practices involve the development and maintenance of relationships with stakeholders, including "value chain actors." Thus, S-LCA and CSR can be integrated for further application.

6 Future of S-LCA-Based CSR Policies

The concept of social sustainability is based on the Universal Declaration of Human Rights given by the United Nations in 1948, which states, "everyone has the right to work, (…) the right to equal pay for equal work, (…) to just and favorable remuneration ensuring for himself and his family an existence worthy of human dignity, and supplemented, if necessary, by other means of social protection (…) has the right to form and to join trade unions for the protection of his interests." To ensure that these ideals enshrined in the Human Rights Declaration are met by the industry, integrating CSR with S-LCA is important.

Amongst the first steps to be undertaken is enhancing the methodologies and indicators used for assessing the social impacts of CSR-integrated S-LCA. This includes exploring innovative approaches, such as incorporating qualitative data, social network analysis, and participatory methods, to capture a more comprehensive understanding of social aspects throughout the life cycle. It is also crucial to explore the synergies and interlinkages between CSR, S-LCA, and other sustainability tools, such as environmental LCA, carbon footprinting, and sustainability reporting frameworks, since the overall objective of a company is sustainable growth. The future of CSR-integrated S-LCA lies in developing approaches for effective stakeholder engagement, ensuring diverse perspectives are considered, and building platforms for dialogue and collaboration among stakeholders to address social challenges collectively. In this respect, CSR policies must be tailored for sector-specific applications. For instance, developing textile sector-specific indicators and guidelines will address the challenges and opportunities of the textile industry as highlighted in Sect. 1. This is aided by understanding how policies and regulations at local, national, and international levels influence meaningful change in CSR.

Effective integration of CSR with S-LCA requires a skilled workforce capable of conducting assessments, interpreting results, and implementing necessary changes. This would require developing educational programs, training modules, and capacity-building initiatives to equip professionals with the knowledge and skills required for successful implementation without additional burden on companies' existing investment, which is a big challenge. But this may be resolved by bigger companies demonstrating the business case and return on investment of CSR-integrated S-LCA initiatives. The demonstrable benefits should be effectively disseminated among researchers, practitioners, policymakers, and civil society through platforms for sharing best practices, case studies, and success stories to get SMEs involved in CSR.

7 Conclusion

The textile and clothing sector has adversely impacted society and the environment on various fronts. The social sustainability-related areas of concern include labor and human rights regulations such as fair wages, safe working conditions, workers' rights to collective bargaining, child labor, forced labor, and discrimination. This impact has been driven by the industry's relocation of production to countries offering cheaper labor, which in turn has spurred a downward spiral of reduced prices, fueling consumer expectations for affordable fashion. However, this shift in textile and clothing manufacturing to middle- or lower-income nations has also promoted their economic development and broadened employment opportunities, particularly for marginalized sections of society. Hence, companies must investigate their social responsibility to ensure that the interests of all stakeholders in the textile supply chain are safeguarded.

To promote sustainability, there is a need to ensure capacity-building programs that provide training on topics such as fair trade, social auditing, and stakeholder engagement. At the individual level, suppliers should be encouraged to follow a code of conduct to promote responsible practices throughout the supply chain. This initiative may be aided by financial incentives such as tax breaks or grants for businesses that adopt socially responsible practices. Where social issues are unclear, businesses can foster partnerships between industry stakeholders, NGOs, and academia to identify social sustainability challenges in the textile industry. The alignment of CSR and S-LCA, as discussed in this chapter, offers a roadmap for responsible decision-making. Integrating S-LCA into CSR frameworks enables a comprehensive understanding of social impacts and supports sustainability in the textile and fashion industry. CSR is either implemented as a separate policy from the business model or is completely integrated into the company's functioning. The latter case allows for more social responsibility in the company's strategies.

Acknowledgments The authors acknowledge the funding received from the European Union's Horizon 2020 research and innovation program under grant agreement No [101060343], Textile Recycling EXcellence (T-REX) project.

Declaration of Generative AI and AI-Assisted Technologies in the Writing Process During the preparation of this work, the author(s) used ChatGPT to improve the readability of certain sections. After using this tool/service, the author(s) reviewed and edited the content as needed and take(s) full responsibility for the content of the publication.

References

Abbate, S., Centobelli, P., Cerchione, R., Nadeem, S. P., & Riccio, E. (2023). Sustainability trends and gaps in the textile, apparel and fashion industries. In *Environment, development and sustainability*. Springer Science and Business Media B.V. https://doi.org/10.1007/s10668-022-02887-2

Aparcana, S., & Salhofer, S. (2013). Application of a methodology for the social life cycle assessment of recycling systems in low income countries: Three Peruvian case studies. *The International Journal of Life Cycle Assessment, 18*(5), 1116–1128. https://doi.org/10.1007/s11367-013-0559-3

Borms, L., Van Opstal, W., Brusselaers, J., & Van Passel, S. (2023). The working future: An analysis of skills needed by circular startups. *Journal of Cleaner Production, 409*, 137261. https://doi.org/10.1016/j.jclepro.2023.137261

BSCI. (2013). *The business social compliance initiative annual report*. https://s3-us-west-2.amazonaws.com/ungc-production/attachments/cop_2014/99871/original/BSCI-2014-05-26-Annual_Report_2013-UK-A4-Mai_2014-Def-Internet.pdf?1406819561.

Bubicz, M. E., Dias Barbosa-Póvoa, A. P. F., & Carvalho, A. (2021). Social sustainability management in the apparel supply chains. *Journal of Cleaner Production, 280*, 124214. https://doi.org/10.1016/j.jclepro.2020.124214

Burhan Wazir. (2001). Nike accused of tolerating sweatshops. *The Guardian*. https://www.theguardian.com/world/2001/may/20/burhanwazir.theobserver

Byrd, K., & Su, J. (2021). Investigating consumer behaviour for environmental, sustainable and social apparel. *International Journal of Clothing Science and Technology, 33*(3), 336–352. https://doi.org/10.1108/IJCST-03-2020-0040

Capasso, P. A. (2021). *Giant's approach to CSR in apparel industry: A comparison between VF corporation and PVH corporation case studies* (issue 717571). https://tesi.luiss.it/32389/1/717571_TRANSI_ILARIA.pdf

Chan, H.-L., Wei, X., Guo, S., & Leung, W.-H. (2020). Corporate social responsibility (CSR) in fashion supply chains: A multi-methodological study. *Transportation Research Part E: Logistics and Transportation Review, 142*, 102063. https://doi.org/10.1016/j.tre.2020.102063

Chawla, J. (2019). *A study of corporate social responsibility (CSR) practices in the Indian textile industry: A pathway to building a sustainable supply chain*. Washington State University. https://hdl.handle.net/2376/102574

Chhipi-Shrestha, G. K., Hewage, K., & Sadiq, R. (2015). 'Socializing' sustainability: A critical review on current development status of social life cycle impact assessment method. *Clean Technologies and Environmental Policy, 17*(3), 579–596. https://doi.org/10.1007/s10098-014-0841-5

Circle Economy. (2023). *Circularity gap report*. https://www.circularity-gap.world/2023

de Abreu, M. C. S. (2015). Perspectives, drivers, and a roadmap for corporate social responsibility in the textile and clothing industry. In S. S. Muthu (Ed.), *Roadmap to sustainable textiles and clothing* (pp. 1–21). Springer. https://doi.org/10.1007/978-981-287-164-0_1

Directorate-General for Environment. (2022). *EU strategy for sustainable and circular textiles*. https://environment.ec.europa.eu/publications/textiles-strategy_en

Dreyer, L., Hauschild, M., & Schierbeck, J. (2006). A framework for social life cycle impact assessment. *The International Journal of Life Cycle Assessment, 11*(2), 88–97. https://doi.org/10.1065/lca2005.08.223

EEA European Environment Agency. (2023). *EU exports of used textiles in Europe's circular economy*. https://www.eea.europa.eu/publications/eu-exports-of-used-textiles

Ekener-Petersen, E., & Moberg, Å. (2013). Potential hotspots identified by social LCA–part 2: Reflections on a study of a complex product. *The International Journal of Life Cycle Assessment, 18*(1), 144–154. https://doi.org/10.1007/s11367-012-0443-6

Ferioli, M., Gazzola, P., Grechi, D., & Vătămănescu, E. M. (2022). Sustainable behaviour of B corps fashion companies during Covid-19: A quantitative economic analysis. *Journal of Cleaner Production, 374*(August). https://doi.org/10.1016/j.jclepro.2022.134010

Fontes, J., Tarne, P., Traverso, M., & Bernstein, P. (2018). Product social impact assessment. *The International Journal of Life Cycle Assessment, 23*(3), 547–555. https://doi.org/10.1007/s11367-016-1125-6

Govindasamy, V., & Suresh, K. (2018). Corporate social responsibility in practice: The case of textile, knitting and garment industries in Malaysia. *Pertanika Journal of Social Sciences and Humanities, 26*(4), 2643–2656.

Grubert, E. (2018). Rigor in social life cycle assessment: Improving the scientific grounding of SLCA. *The International Journal of Life Cycle Assessment, 23*(3), 481–491. https://doi.org/10.1007/s11367-016-1117-6

Herrera Almanza, A. M., & Corona, B. (2020). Using social life cycle assessment to analyze the contribution of products to the sustainable development goals: A case study in the textile sector. *International Journal of Life Cycle Assessment, 25*(9), 1833–1845. https://doi.org/10.1007/s11367-020-01789-7

International Labor Organization. (2023). *How to achieve gender equality in global garment supply chains*. International Labor Organization. https://www.ilo.org/infostories/en-GB/Stories/discrimination/garment-gender#introduction

ISO. (2006a). ISO 14040: Environmental management – life cycle assessment. *Principles and framework*.

ISO. (2006b). ISO 14044: Environmental management – life cycle assessment. *Requirements and guidelines*.

Judd, J., & Kuruvilla, S. (2020, April 30). Why apparel brands' efforts to police their supply chains aren't working. *The Conversation*. https://theconversation.com/why-apparel-brands-efforts-to-police-their-supply-chains-arent-working-136821

Khattak, A., Haworth, N., Stringer, C., & Benson-Rea, M. (2017). Is social upgrading occurring in South Asia's apparel industry? *Critical Perspectives on International Business, 13*(3), 226–243. https://doi.org/10.1108/cpoib-11-2015-0051

Księżak, P. (2017). The CSR challenges in the clothing industry. *Journal of Corporate Responsibility and Leadership, 3*(2), 51. https://doi.org/10.12775/jcrl.2016.008

Kumar, A., Moktadir, A., Liman, Z. R., Gunasekaran, A., Hegemann, K., & Rehman Khan, S. A. (2020). Evaluating sustainable drivers for social responsibility in the context of ready-made garments supply chain. *Journal of Cleaner Production, 248*, 119231. https://doi.org/10.1016/j.jclepro.2019.119231

Kumari, A., & Singh, M. P. (2023). A journey of social sustainability in organization during MDG & SDG period: A bibliometric analysis. *Socio-Economic Planning Sciences, 88*. https://doi.org/10.1016/j.seps.2023.101668

Lenzo, P., Traverso, M., Salomone, R., & Ioppolo, G. (2017). Social life cycle assessment in the textile sector: An Italian case study. *Sustainability, 9*(11), 2092. https://doi.org/10.3390/su9112092

Mahiat, T., Al Alam, M. A., Argho, M., Corlett, J., Chowdhury, R. B., Biswas, K. F., Hossain, M. M., & Sujauddin, M. (2022). Modeling the environmental and social impacts of the handloom industry in Bangladesh through life cycle assessment. *Modeling Earth Systems and Environment, 0123456789*. https://doi.org/10.1007/s40808-022-01491-7

María, A., Almanza, H., Corona, B., Herrera Almanza, A. M., & Corona, B. (2020). Using social life cycle assessment to analyze the contribution of products to the sustainable development goals: A case study in the textile sector. *International Journal of Life Cycle Assessment, 25*(9), 1833–1845. https://doi.org/10.1007/s11367-020-01789-7

Martin, M., & Herlaar, S. (2021). Environmental and social performance of valorizing waste wool for sweater production. *Sustainable Production and Consumption, 25*, 425–438. https://doi.org/10.1016/j.spc.2020.11.023

McKinsey & Company. (2022). *Scaling textile recycling in Europe-turning waste into value.*. https://vb.nweurope.eu/media/19019/2207-scaling-textile-recycling-in-europe-turning-waste-into-value.pdf

McDonough, W., & Braungart, M. (2002). *Cradle to Cradle: Remarking the Way We Make Things*. North Point Press, New York.

Meier, C. (2015). Supply chains in the apparel industry: Do transnational initiatives for social sustainability improve workers' situation? *International Journal of Management, Knowledge and Learning, 4*, 27–40. https://api.semanticscholar.org/CorpusID:154491248

Mukendi, A., Davies, I., Glozer, S., & McDonagh, P. (2020). Sustainable fashion: Current and future research directions. *European Journal of Marketing, 54*(11), 2873–2909. https://doi.org/10.1108/EJM-02-2019-0132

Muñoz-Torres, M. J., Fernández-Izquierdo, M. Á., Ferrero-Ferrero, I., Escrig-Olmedo, E., & Rivera-Lirio, J. M. (2023). Social life cycle analysis of textile industry impacts for greater social sustainability of global supply chains. *System, 11*(1). https://doi.org/10.3390/systems11010008

Niinimäki, K. (2018). Sustainable fashion in a circular Economy. In K. Niinimäki (Ed.), *Sustainable fashion in a circular Economy* (pp. 12–41). Aalto ARTS Books. http://urn.fi/URN:ISBN:978-952-60-0090-9

Niinimäki, K. (2023). New fashion ethics: Who has the justice and value in fashion? In A.-M. Almila & D. S. Almila (Eds.), *Fashion's transnational inequalities: Socio-political, economic, environmental* (pp. 83–99). Routledge. https://doi.org/10.4324/9781003219675-5

Niinimäki, K., & Karell, E. (2020). Closing the loop: Intentional fashion design defined by recycling technologies. In G. Vignali, L. F. Reid, D. Ryding, & C. E. Henninger (Eds.), *Technology-driven sustainability: Innovation in the fashion supply chain* (pp. 7–25). Springer International Publishing. https://doi.org/10.1007/978-3-030-15483-7_2

Papo, M., & Corona, B. (2022). Life cycle sustainability assessment of non-beverage bottles made of recycled high density polyethylene. *Journal of Cleaner Production, 378*, 134442. https://doi.org/10.1016/j.jclepro.2022.134442

Potting, J., Hekkert, M. P., Worrell, E., & Hanemaaijer, A. (2017). *Circular Economy: Measuring innovation in the product chain*. PBL Netherlands Assessment Agency.

Roos, S., Zamani, B., Sandin, G., Peters, G. M., & Svanström, M. (2016). A life cycle assessment (LCA)-based approach to guiding an industry sector towards sustainability: The case of the Swedish apparel sector. *Journal of Cleaner Production, 133*, 691–700. https://doi.org/10.1016/j.jclepro.2016.05.146

Saxena, I. (2021). Is mandatory CSR a good idea for textile & apparel industry? *Apparel Resources*. https://in.apparelresources.com/business-news/sustainability/mandatory-csr-good-idea-textile-apparel-industry/

Shaikh, S., Thomas, K., Zuhair, S., & Magalini, F. (2020). A cost-benefit analysis of the downstream impacts of e-waste recycling in Pakistan. *Waste Management, 118*, 302–312. https://doi.org/10.1016/j.wasman.2020.08.039

Staicu, D., & Pop, O. (2018). Mapping the interactions between the stakeholders of the circular economy ecosystem applied to the textile and apparel sector in Romania. *Management & Marketing, 13*(4), 1190–1209. https://doi.org/10.2478/mmcks-2018-0031

Su, J., Watchravesringkan, K., & (Tu), Zhou, J., & Gil, M. (2019). Sustainable clothing: Perspectives from US and Chinese young millennials. *International Journal of Retail & Distribution Management, 47*(11), 1141–1162. https://doi.org/10.1108/IJRDM-09-2017-0184

Suarez-Visbal, L. J., Carreón, J. R., Corona, B., & Worrell, E. (2022). The social impacts of circular strategies in the apparel value chain; a comparative study between three countries. *Circular Economy and Sustainability* (issue 0123456789). https://doi.org/10.1007/s43615-022-00203-8

Szewczyk, M. (2016). Corporate social responsibility in the textile and apparel industry: Barriers and challenges. *Journal of Corporate Responsibility and Leadership, 3*(4). https://doi.org/10.12775/JCRL.2016.024

The Observatory of Economic Complexity. (2021). *Textiles in Haiti*. https://oec.world/en/profile/bilateral-product/textiles/reporter/hti

Thorisdottir, T. S., & Johannsdottir, L. (2020). Corporate social responsibility influencing sustainability within the fashion industry. A systematic review. *Sustainability, 12*(21), 9167. https://doi.org/10.3390/su12219167

Traverso, M., Mankaa, M. N., Valdivia, S., Roche, L., Luthin, A., Garrido, S. R., & Neugebauer, S. (2022). Pilot Projects on Guidelines for Social Life Cycle Assessment of Products and Organizations.

UNEP. (2020). *Guidelines for social life cycle assessment of products and organisations 2020.* https://wedocs.unep.org/20.500.11822/34554

UNEP. (2021). In M. Traverso, S. Valdivia, A. Luthin, L. Roche, G. Arcese, S. Neugebauer, L. Petti, M. D'Eusanio, B. M. Tragnone, R. Mankaa, J. Hanafi, C. B. Norris, & A. Zamagni (Eds.), *Methodological sheets for subcategories in Social Life Cycle Assessment (S-LCA).*

van der Velden, N. M., & Vogtländer, J. G. (2017). Monetisation of external socio-economic costs of industrial production: A social-LCA-based case of clothing production. *Journal of Cleaner Production, 153*, 320–330. https://doi.org/10.1016/j.jclepro.2017.03.161

Walker, A. M., Opferkuch, K., Roos Lindgreen, E., Simboli, A., Vermeulen, W. J. V., & Raggi, A. (2021). Assessing the social sustainability of circular economy practices: Industry perspectives from Italy and The Netherlands. *Sustainable Production and Consumption, 27*, 831–844. https://doi.org/10.1016/j.spc.2021.01.030

World Trade Organization. (2022). *World trade statistical review.* https://www.wto.org/english/res_e/booksp_e/wtsr_2022_e.pdf

Yıldız-Geyhan, E., Altun-Çiftçioğlu, G. A., & Kadırgan, M. A. N. (2017). Social life cycle assessment of different packaging waste collection system. *Resources, Conservation and Recycling, 124*, 1–12. https://doi.org/10.1016/j.resconrec.2017.04.003

Zamani, B., Sandin, G., Svanström, M., & Peters, G. M. (2018). Hotspot identification in the clothing industry using social life cycle assessment – Opportunities and challenges of input-output modelling. *The International Journal of Life Cycle Assessment, 23*(3), 536–546. https://doi.org/10.1007/s11367-016-1113-x

Zerbino, P., Stefanini, A., Aloini, D., Dulmin, R., & Mininno, V. (2023). Does supply chain sustainability benefit from formal scavenging? A case study in circular settings. *Journal of Cleaner Production, 385*(December 2022), 135669–135669. https://doi.org/10.1016/j.jclepro.2022.135669

[illegible] (2020). [illegible] and their environmental [illegible]. [illegible] (2021), 9145. https://doi.org/10.[illegible]

[illegible] (2021). [illegible]. Department of [illegible].

[illegible] (2019). [illegible]. https://doi.org/10.[illegible]

[illegible], M., [illegible], A., [illegible], L., [illegible], K., [illegible] (2019). [illegible]. 113.

[illegible] (2019). [illegible] of external cost [illegible] of [illegible] of clothing production [illegible]. [illegible], 2(2), 150. https://doi.org/10.[illegible]

Wallace, M., [illegible] (2021). [illegible] lifecycle [illegible] of clothing [illegible]. *[illegible] Journal of [illegible]*, 22, 1-[illegible]. https://doi.org/10.[illegible]

World [illegible] (2022). [illegible]. https://www.[illegible]

[illegible] (2017). [illegible] different [illegible] washing [illegible]. *Resources, Conservation and Recycling*, [illegible]. https://doi.org/10.1016/j.resconrec.2017.[illegible]

[illegible] (2018). [illegible] identification in [illegible] the fashion industry [illegible] opportunities and challenges of [illegible]. *[illegible]*, 13(4), 153–[illegible]. https://doi.org/10.[illegible]

[illegible] (2023). Does [illegible] sustainability [illegible] in [illegible]? *Journal of Cleaner Production*, 413, 137508. https://doi.org/10.1016/j.jclepro.2023.137508

CSR and Commitment to the Slow Fashion: The Adolfo Dominguez Case Study

M. Isabel Sánchez-Hernández and Lara Bastos-Sánchez

Abstract The textile industry has historically faced various challenges related to social and environmental sustainability. Many textile businesses are adopting corporate social responsibility (CSR) initiatives not only as a moral imperative but also as a strategic business move to meet evolving consumer expectations and enhance their brand reputation. The intersection of slow fashion and CSR is particularly relevant in addressing these concerns, emphasizing the importance of sustainable and ethical practices in the production and consumption of textiles. In this context, this chapter presents the experience of the company Adolfo Dominguez, a Spanish fashion firm that has gained relevance after showing its strong commitment to a more environmentally friendly consumer model. In one of the first collections of the firm, the slogan "The wrinkle is beautiful" marked a milestone in Spanish advertising. It is a concept that the brand has recalled and reused in recent campaigns, along with "Be Older" to promote the sustainability in the garments of its collections and to join the slow-fashion concept. The objective is to avoid wasting resources by buying quality garments and fighting against fast fashion. First, this chapter offers a theoretical framework defending the need to develop CSR strategies in the textiles and fashion industry. Second, the chapter presents the case study of the Adolfo Dominguez Spanish firm, with specific attention to its impactful marketing campaigns and its manifesto for animal welfare, for a more conscious and responsible consumption model, or the last initiative focused on reuse that allows consumers to rent, instead of to buy, clothing, and accessories. Finally, comparing the firm to competitors, the conclusion reflects on the new sustainable trends in the fashion industry.

M. I. Sánchez-Hernández (✉)
University of Extremadura, Badajoz, Spain
e-mail: isanchez@unex.es

L. Bastos-Sánchez
KEDGE Business School, Paris, France
e-mail: lara.bastossanchez@kedgebs.com

S. S. Muthu (ed.), *Corporate Social Responsibility in Textiles and Fashion*,
Sustainable Textiles: Production, Processing, Manufacturing & Chemistry,
https://doi.org/10.1007/978-3-031-61099-8_6

Keywords CSR · Circular economy · Fashion · Fast fashion · Slow fashion · Spain · Sustainability · Textile industry

1 Introduction

Fashion serves as a reflection of the prevailing taste, dominant values, and societal perspectives of a given era. It functions as both a mirror and an aesthetic manifestation of a particular society, providing valuable insights into the prevailing mindset of the time. Throughout different historical epochs, fashion trends are not solely shaped by prominent figures such as artists and politicians; rather, they are intrinsically tied to the prevailing values and social attitudes of the period. This is because fashion is a form of visual communication (Lee et al., 2018) and acts as a medium for conveying identity, especially now in the twenty-first century (Vänskä & Gurova, 2022). Presently, there is a growing awareness within society regarding environmental sustainability. Within this context, it is imperative to emphasize the role of corporate social responsibility (CSR) about environmental concerns, active engagement, and awareness among consumers and about the public sector's responsibility in fostering wealth creation.

When directing our attention toward the consumer, it becomes evident that a distinct consumer segment is characterized by a profound concern for environmental preservation and the well-being of the planet. This group may aptly be denominated as the ecological consumer or the green consumer. This segment is formed by individuals who exhibit a set of distinctive characteristics and behaviors centered around environmental sustainability and ethical consumption. They are highly aware of environmental issues, including climate change, pollution, resource depletion, and habitat destruction. They view their consumption decisions as a form of personal responsibility in the global effort to address environmental challenges.

In general terms, they aim to reduce waste and minimize the consumption of disposable and non-recyclable products and often favor locally sourced goods to reduce the carbon footprint associated with transportation and promote local economies. Thus, they are inclined to repair, repurpose, or buy second-hand items, extend the lifespan of products, and reduce waste. In terms of price, they are generally willing to pay a premium for products and services that align with their environmental and ethical values.

In the textile industry, these characteristics reflect the values and behaviors of ecological or green consumers who are actively seeking ways to align their lifestyles with a commitment to environmental sustainability and ethical consumption. In this chapter we present and discuss the marketing campaigns of the international company Adolfo Dominguez SA, a Spanish fashion firm that has gained relevance after showing its strong commitment to a more environmentally friendly consumer model.

2 Theoretical Framework: From Fast to Slow Fashion

The fashion sector is widely recognized as one of the principal industries responsible for exerting a significant impact on environmental degradation (Grazzini et al., 2021). The fashion industry is commonly regarded as lacking sustainability primarily due to the proliferation of fast fashion (Adamkiewicz et al., 2022). The term *fast fashion industry* refers to a business model and approach in the fashion industry that emphasizes rapid production and delivery of inexpensive, trendy clothing to meet consumer demand. Fast fashion brands prioritize quick design-to-production cycles, often copying popular styles from the runway or other fashion influencers and making them available to consumers at a much lower cost. According to Joy et al. (2012), fast fashion mimics current luxury fashion collections for young consumers with low purchasing power.

The fast fashion has gained prominence over the last few decades and is characterized by its ability to respond rapidly to changing fashion trends and to deliver new collections to stores and online retailers on a frequent basis. The academic literature offers a comprehensive understanding of the fast fashion industry, its origins, challenges, and consumer behavior (Barnes & Lea-Greenwood, 2006; Caro & Martínez-de-Albéniz, 2015; Todeschini et al., 2017; Liu et al., 2020) but also its impact on sustainability (Niinimäki et al., 2020; Peters et al., 2021). As a result, nowadays there is a need to develop corporate social responsibility (CSR) and circular economy strategies in the textile and fashion industry (Abbate et al., 2023).

An illustration of the significance of sustainability in the textile industry is the endeavors of international companies such as the Inditex Group. These companies have faced criticism for their fast fashion business model. However, simultaneously, they make efforts to communicate their sustainability practices and engage in responsible actions, including corporate donations (Galán-Ladero & Sánchez-Hernández, 2022).

An increasing scholarly focus has emerged on the concept of *slow fashion*, which denotes a production methodology accentuating the primacy of quality as a means to attain sustainability within the fashion industry (Jung & Jin, 2016). It has been said that the *slow fashion* is a reaction to the *fast fashion* and the answer of the sector for a sustainable fashion industry (Štefko & Steffek, 2018).

3 Method

According to Quiles-Soler et al. (2023), fashion brands use social media and corporate websites as vehicles for communicating ethics, CSR, and sustainability. We have conducted the exploratory descriptive case study of the Spanish fashion brand Adolfo Dominguez, focused on its marketing campaigns, with the aim of attaining an in-depth understanding of the content of such communications related to the slow-fashion commitment by the company, and their relationship to sustainability.

Later, we compare the company's message of sustainability to competitors to corroborate its leadership as a slow-fashion textile firm.

The rest of the chapter presents the case study of Adolfo Dominguez SA, the company, the general commitment to sustainability, and the analysis of a selection of campaigns illustrating the business case for sustainability. We finish with the analysis of the alternative actions related to sustainability from competitive firms, and a general discussion, to conclude imagining the future.

4 Case Study: Adolfo Dominguez SA

4.1 The Company

Adolfo Dominguez SA stands as a prominent and highly recognized fashion conglomerate in Spain. The company is engaged in the comprehensive design and manufacture of clothing targeting diverse segments, including men's, women's, and youth fashion. The firm markets its apparel through various brands such as AD and Linea U. Beyond its primary focus on clothing, Adolfo Dominguez extends its product portfolio to encompass a wide array of accessory items and footwear, a substantial portion of which is crafted under licensing agreements with third-party entities. Moreover, the company has ventured into the production of its own line of fragrances to complement its primary labels, a venture realized in collaboration with the fragrance specialist, Myrurgia.

Adolfo Dominguez SA is a company with a rich history. Some key milestones from its origins to the present are the following:

- Founding (1973): Adolfo Dominguez SA was established in Ourense, Spain, by Adolfo Dominguez, starting as a small boutique. Its position as a vertically integrated company has enabled it to produce high-quality, fashion-oriented clothing at attractive price points. The company's vertical integration extended into the retail sector.
- Expansion (1980s): In the 1980s, the company began expanding its operations internationally, opening stores in other European countries.
- Initial public offering (1997): The company went public and was listed on the Spanish stock exchange, allowing it to access capital for further growth and development.
- Diversification (2000s): Adolfo Dominguez diversified its product lines, introducing accessories, fragrances, and footwear in addition to its clothing collections. The company also experimented with home furnishings.
- Sustainability focus (2010s): The company started to place a strong emphasis on sustainability and ethical fashion. Initiatives were launched to reduce its environmental impact and promote responsible practices within the fashion industry.
- Retail expansion: Adolfo Dominguez continued to expand its retail presence, with stores in major cities in Spain and internationally. Nowadays the principal

subsidiaries are the following: Adolfo Dominguez Arg. SA (Argentina), Adolfo Dominguez Belgique, Adolfo Dominguez GmbH (Germany), Adolfo Dominguez Japan Ltd., Adolfo Dominguez Ltd. (United Kingdom), Adolfo Dominguez Lux. SA (Luxembourg), Adolfo Dominguez Moda Ltd. (Portugal), Adolfo Dominguez SARL (France), Adolfo Dominguez USA Inc., and Trespass SA de CV (Mexico).
- Collaborations and special collections: The brand engaged in collaborations with artists and designers and introduced special collections, contributing to its brand identity.
- Leadership transitions: Leadership changes occurred as Adolfo Dominguez stepped down from the role of CEO, and his daughter, Adriana Dominguez, assumed a leadership role within the company.
- Digital transformation: Adolfo Dominguez embraced e-commerce and digital marketing strategies to reach a broader online audience.
- COVID-19 impact: Like many fashion companies, Adolfo Dominguez had to adapt to the challenges posed by the COVID-19 pandemic, including store closures and changes in consumer behavior.

The present is promising according to data offered by the company in its web page.[1] The Adolfo Dominguez group witnessed a remarkable 24% growth in sales during the 2022–2023 fiscal year, spanning from March 2022 to February 2023. This impressive expansion translated into a substantial turnover of 114.2 million euros. Notably, the company's financial performance exhibited a noteworthy enhancement in profitability, as evidenced by the significant rise in the gross margin to 56.8%. This surge in profitability culminated in the achievement of a net profit of 0.2 million euros, following a positive trajectory from a pre-tax figure of −1.3 million euros. Of equal importance is the company's remarkable success in bolstering its EBITDA (earnings before interest, taxes, depreciation, and amortization). In comparison to the previous year, the firm succeeded in multiplying its EBITDA by a remarkable 35-fold, resulting in an impressive operating result of 13.4 million euros, which marked a substantial upswing from the modest 0.4 million euros recorded just a year ago.

The brand's sales are experiencing robust growth across all the global markets in which it maintains a presence. Particularly noteworthy is the remarkable surge of 30.6% in sales outside of the European market, with Mexico notably standing out with a remarkable increase of 44.7%, marked by the establishment of 128 new stores. Simultaneously, the European market witnessed substantial year-on-year sales growth, recording a significant upswing of 21.7%.

The company's substantial expansion into foreign markets, which has been notably accelerated in recent years, now constitutes a significant portion, accounting for 57% of its global retail network. Over the course of the year, Adolfo Dominguez inaugurated 27 new stores, with a majority—two-thirds—of these establishments located beyond Spain's borders. Furthermore, the company recently inked a pivotal agreement with the esteemed Chalhoub Group, heralding a substantial fivefold

[1] Adolfo Dominguez Web Page https://www.adolfodominguez.com [Accessed 20/09/2023]

expansion in its footprint within the Middle East. This agreement encompasses the launch of 16 new stores across the United Arab Emirates, Saudi Arabia, Egypt, Jordan, Lebanon, and Bahrain.

Online sales exhibit a resilient growth rate of 2.2% when compared to the preceding year, demonstrating their sustained vigor even following the complete resumption of in-store foot traffic in the post-COVID period. Notably, in recent months, the brand has embarked on a strategic expansion, introducing its range of clothing and accessories to prominent online marketplaces, including Zalando and About You.

As of the present moment, the Adolfo Dominguez group boasts a workforce of 957 dedicated professionals and a robust retail presence comprising 339 points of sale spanning across 18 countries by the year's end. Noteworthy is the composition of the professional staff, where 80% of employees are women, and the management committee, in particular, commands a substantial 63% representation of executives.

4.2 The Ethical and Value-Based Commitment to Sustainability of the Company

As we have seen in previous sections, fast fashion has been taking over the textile industry, with precarious working conditions and devastating consequences for the environment. However, more and more brands are joining in to stop this event and offer alternatives with optimal quality standards, good manufacturing conditions, and environmentally friendly resources. In this case, Adolfo Dominguez has been launching different collections for a few years that precisely promote awareness about the phenomenon of fast fashion and the need to reuse clothing to prevent waste and pollution generated in the environment.

Since its origins, Adolfo Dominguez's ethics and sustainability values have been reflected in strategic plans focused on having a positive impact. The strategic plans are oriented toward the ongoing enhancement of all organizational processes, aiming to foster collaboration with sustainable development, adherence to legal obligations, a transparent and unequivocal dedication to proactive measures, and the establishment of control mechanisms designed to mitigate any risks that may impact both society and the environment.

From this, we have analyzed some of the main communications that are carried out after the launch of sustainable collections and that have two clear objectives: on the one hand, to promote a culture of awareness about sustainability in the textile industry and, on the other hand, to position Adolfo Dominguez as a national author brand that carries out actions in accordance with this ethic and sustainable philosophy and values. The campaigns have been carried out at the offline marketing level, with media distributed in different cities, in press, and in the physical establishments themselves, as well as online and communications developed on social networks such as Instagram or YouTube and on its own website.

For instance, animal welfare falls within the ethics and values of Adolfo Dominguez. In 2010 the company spoke out against the overuse of leather, creating an animal welfare policy, gradually promoting alternatives to leather and fur, and their combination with more vegan options. Thus, Adolfo Dominguez can be defined in a few words as signature garments with an elegant and comfortable design combined with quality and durability that guarantee sustainability and curb the use of resources.

In relation to the Sustainable Development Goals (SDGs) approved by the UN in 2015, Adolfo Dominguez works to identify those SDGs that are most directly linked to the actions carried out by the company and that are described throughout this document and studies how to integrate them into the Positive Impact Strategic Plan to turn them into specific, measurable, achievable, realistic, and time-bound objectives that reflect your responsible commitment. Among some initiatives worth highlighting in this area would be internal training through communications and Intranet on the importance of the SDGs and internal awareness of them. So much so, in 2018 the Spanish Network of the United Nations Global Compact, on the third anniversary of the SDGs, selected the campaign that we will analyze later, "Be Older," as an example of good practice for SDG 12—Responsible production and consumption.

Recently, the firm has focused on reuse and allows you to rent clothing and accessories for 4 or 8 days including home delivery and collection. The project is based on the brand's efforts for more conscious fashion consumption.

4.3 Marketing Campaigns and Its Relationship to Sustainability

Since its inception, the brand's communication had been underpinned by an emphasis on quality and durability in contrast to the transience of fast fashion. When the company elects to strategically address sustainability and places it at the core of its communication, it embarks on advertising campaigns that delve into the notion of timelessness in design, aiming to tackle the endemic issue of season acceleration and fashion with built-in obsolescence. We can affirm that Adolfo Dominguez has an innovative ideology with regard to fashion communication, based on clear messages in a direct and honest dialogue with a conscious consumer as we will show in the following lines.

In the 1980s, in 1983, one of the first collections of the firm had the slogan "The wrinkle is beautiful." This campaign marked a milestone in Spanish advertising and had high recognition in international markets. This notorious expression alluded not just to the qualities of linen but also to the natural styles and the belief that fashion serves as a metaphorical epidermis or second skin. Forty years later, the company is a benchmark for global sustainable fashion.

Adolfo Dominguez perceives fashion as a reflection of society, hence his alignment with the global awakening of consciousness through the anti-fur or fur-free movement, wherein fashion brands express their rejection of the use of animal skins and animal cruelty. Instead, they substitute these materials with synthetic furs and other technological materials, under the motto: "the new luxury that represents us." In 2010, and after regular correspondence with People for the Ethical Treatment of Animals (PETA),[2] the largest animal rights organization in the world, the company decided to introduce a comprehensive and progressive new animal welfare policy, which included a commitment to use only "nonmuled" or "mulesing-free" wood and ban fur and exotic skins.

In 2020, the campaign titled "Old Clothes" resulted in a remarkable increase of 32.76% in net sales in comparison to the corresponding period of 2019. Moreover, it attracted 27.39% more new visitors to its website in contrast to the figures from 2019. This initiative was framed within their overarching strategy known as "Be Older" (Fig. 1) through which the advertiser aimed to confront excessive fast-fashion consumption by advocating for a more rational approach to shopping. In this campaign, Adolfo Dominguez invited consumers to reflect on the quality and durability of garments in a system that increasingly has similarities with fast fashion. "Old people are not impulsive, nor capricious, before doing something, they think about it […] Think if it suits you. Think if it will last. Think if you need it. Think. Then Buy," they recognized from the brand. Consequently, the brand emerged as one of the most active proponents of sustainability in the fashion industry. The campaign championed the idea that designers should create durable clothing, thereby contributing to the reduction of fashion's status as the second most environmentally polluting industry.

The clear marketing strategy carried out by the firm was to rebel against the current consumption model that characterizes the textile industry, where a more

Source: Adolfo Dominguez web page (https://adolfodominguez.com) [18/10/2023]

Fig. 1 Visual excerpts from the "Be Older" campaign. (*Source:* Adolfo Dominguez web page (https://adolfodominguez.com) [18/10/2023])

[2] PETA entities have more than nine million members and supporters globally. Web Page: https://www.peta.org [Accessed 11/10/2023]

conscious and responsible one is proposed, in which being "old" serves as wisdom and guidance. With these campaigns, the brand insists that "Old people know things." They have seen a lot of things, and they know that not everything new is necessarily better. That there is something absurd about buying something and not using it. You don't have to buy more but rather choose better. That, if something is perfect, you should repeat it. What is sustainable is to have a skirt that lasts 10 years. That it is better to know about style than about fashion. Therefore, this is a nod to one of the first collections that Adolfo Dominguez launched under the motto "The wrinkle is beautiful" but adapted to the values that the brand wants to communicate: quality garments, good finishes, and timelessness.

In pursuit of this objective, an unprecedented endeavor was undertaken within the realm of the fashion industry: a campaign featuring clothing from past decades. Under the appellation "Vintage Attire," a total of 29 garments, some of which dated back up to 40 years, were showcased. These garments were generously contributed by real customers who not only preserved them but also continued to incorporate them into their wardrobes. A photoshoot session was orchestrated with international models, during which the clothing donors were graciously extended an invitation. They actively participated in a series of testimonial videos. Subsequently, a set of visual graphics was conceived, each headline denoting the respective years of origin for the garments, alongside a message urging a fashion consumption imbued with prudence: "Purchase clothing that endures beyond passing trends. Embrace the wisdom of age."

Marketing News (2023),[3] an information medium aimed at marketing professionals, highlights that, to inaugurate the campaign, an elderly actress from Galicia, the Spanish region where the company was founded and clearly associated with the brand, attended the Goya Film Awards wearing an ensemble crafted by Adolfo Dominguez from three decades prior. This gesture garnered substantial media coverage and initiated a discourse on the intrinsic value of the vintage. As a culmination, the "Vintage Attire" campaign included the release of a limited collection, reissuing several of the donated garments. This reissue was executed using leftover fabrics to minimize the environmental footprint. In a tribute to the original clothing donors, each piece bore the name of its original owner.

- "Repeat More. Think Less"
- In this campaign (Fig. 2), the firm invites the consumer *thinking less* because this makes us freer. Humans make too many meaningless decisions in our daily lives, and the company invites us to discover the new mental sustainability. Although some people could think that repeating means having no style, according to this campaign, the truth is that repeating is the best way to find our own. It was added: Repeat more. Need less.
- "Think, Then Buy"

[3] Marketing News Web Page: https://www.marketingnews.es/ [Accessed 16/10/2023]

Source: Adolfo Dominguez web page (https://adolfodominguez.com) [18/10/2023]

Fig. 2 Visual excerpts from the "Repeat more. Think less" campaign. (*Source:* Adolfo Dominguez web page (https://adolfodominguez.com) [18/10/2023])

Source: Adolfo Dominguez web page (https://adolfodominguez.com) [18/10/2023]

Fig. 3 Visual excerpts from the "Think, then buy" campaign. (*Source:* Adolfo Dominguez web page (https://adolfodominguez.com) [18/10/2023])

- The "Think, then buy" campaign (Fig. 3) was the continuation of the Be Older initiative. Adolfo Dominguez created a campaign advocating responsible consumption. Playing with Descartes' axiom "I think, therefore I am," the Spanish firm launched a prerogative, "Think, then buy," that invites you to think. Yes, think before waiting in so many lines at the stores. To think first if we really need that garment that we have been craving. Let's think about whether it will soon be part of that volume of clothes that we will end up throwing away when we change our wardrobe. The sculpture of The Thinker, by Frenchman Auguste Rodin, served as the starting point for a very powerful graphic image in which several people appear abstracted in the same posture on piles of discarded clothes.
- "ADN Rent"
- In line with the desire to promote a more rational and conscious fashion consumption, Adolfo Dominguez has launched its own clothing rental service under the name ADN Rent (Fig. 4). The project, which is now open to the public, seeks to offer alternatives to excessive clothing purchasing.

Source: Adolfo Dominguez web page (https://www.adolfodominguez.com/en-fr/se-mas-viejo.html) [18/10/2023]

Fig. 4 Visual excerpts from the "ADN Rent" campaign. (*Source:* Adolfo Dominguez web page (https://www.adolfodominguez.com/en-fr/se-mas-viejo.html) [18/10/2023])

The service has been launched to coincide with the premiere of the brand's autumn-winter 2023 collection, whose designs are available to rent. ADN Rent will have a selection of men's and women's collections that will be renewed season by season. Prices vary depending on the garments: from 130 euros for a men's suit, from 40 euros for a bag, or from 85 euros for a women's jumpsuit. Clothing and accessories can be rented for 4 or 8 days through the website. Users can choose the clothes they want and the days they would like to wear them and can return them within the first 24 h if the size is incorrect. This possibility also includes the shipping of rented items and their collection at home.

After a press release from the fashion company, the communication company Reason Why (2023) highlights what was explained by Patricia Alonso, Corporate Director of Marketing and Communication at Adolfo Dominguez, in the following statements clearly aligned to sustainability:

1. "Dressing ourselves is a functional act, but also an emotional one."
2. "Fashion is the illusion of choosing and wearing a garment for a special event."
3. "We want the joy of wearing a new garment to remain present, to be accessible. And make it possible without the closet filling up with clothes that you only wore one day."

Both the 2020 ADN Box project and this 2023 ADN Rent project are part of the ADN ecosystem, which also includes ADN Premium, a subscription model that provides access to discounts, or ADN Live, which facilitates access to the brand's fashion shows and events. All of them complement the brand's proposal, based on a philosophy that is committed to the design of durable and timeless garments and the creation of more efficient wardrobes.

5 Competitor Analysis

According to Knoji (2023),[4] a shopping knowledge hub where over three million monthly shoppers research before they buy, Adolfo Dominguez is a designer and luxury fashion brand. Per Craft (2023),[5] a prominent US supplier intelligence firm specializing in facilitating data-driven corporate decision-making, Adolfo Dominguez is a company actively involved in the conception and design, production, and retailing of ready-made apparel. Its primary competitors in this sector include the following:

- *Timberland*,[6] a manufacturer and retailer of outdoor wear with a focus on footwear.
- *Anatomie*,[7] a women's travel wear company that specializes in creating luxury travel clothing for the modern woman on-the-go.
- *Love Culture*,[8] considered the latest trends in junior clothing and teen clothing at affordable prices.
- *Duluth Trading Company*,[9] a lifestyle brand of men's and women's casual wear, workwear, and accessories.

Some of the main Spanish companies considered as competitors of Adolfo Dominguez in the global market at this time included:

- *Inditex Group—Zara* is one of the largest and most well-known fast-fashion retailers globally and competes directly with Adolfo Dominguez. *Massimo Dutti* is also part of the Inditex Group and focuses on high-quality clothing and accessories.
- *Mango*—It is another popular Spanish fashion brand with a global presence, known for its contemporary and stylish clothing for women, men, and children.
- *Desigual*—A Spanish clothing brand known for its colorful and distinctive designs.
- *Cortefiel*—A Spanish fashion retailer that caters to a mature and contemporary audience.

Finally, other fashion brands competing with Adolfo Dominguez in international markets are:

- *H&M*—A Swedish multinational fashion retailer known for its affordable and trendy clothing.

---

[4] Knoji Web Page: https://knoji.com/ [Accessed 14/10/2023]

[5] Craf Web Page: https://craft.co/adolfo-dominguez-s-a/competitors [Accessed 20/10/2023]

[6] Timberland Web Page: http://www.timberland.com [Accessed 24/10/2023]

[7] Anatomie Web Page: https://anatomie.com [Accessed 22/10/2023]

[8] Love Culture Web Page: http://www.loveculture.com [Accessed 23/10/2023]

[9] Duluth Trading Company Web Page: https://www.duluthtrading.com/ [Accessed 16/10/2023]

- *Benetton*—An Italian clothing brand that offers a wide range of clothing and accessories.

In line to the main elements of the sustainability strategy of Adolfo Dominguez, and to carry out a comparative study of the sustainability of the selected companies, we have focused our attention on three specific aspects: animal welfare, reuse/recycle, and rent. Results are shown in Table 1, followed by a resume of the main sustainability practices company by company.

Adolfo Dominguez, the Spanish fashion company analyzed in this chapter as case-study, is recognized for its strong commitment to sustainability. They prioritize eco-friendly materials, including organic and sustainable fabrics, and promote responsible production processes, with an emphasis on minimizing waste and water usage. Adolfo Dominguez also implements eco-efficient store designs and transportation strategies to reduce their environmental impact. They emphasize the importance of local and ethical production, supporting fair labor practices. Moreover, the company promotes circular fashion by offering recycling programs and encouraging longevity in their designs. Adolfo Dominguez's comprehensive sustainable practices reflect their dedication to environmental and ethical responsibility in the fashion industry.

Timberland has implemented several key sustainable practices to reduce its environmental footprint and promote responsible production. These practices include the use of eco-friendly materials, such as recycled plastics and organic cotton, in their products, a commitment to reducing water and energy consumption in manufacturing processes, and a dedication to responsible sourcing by supporting fair labor practices and environmentally friendly leather tanning. Timberland also actively engages in reforestation initiatives, with a goal to plant ten million trees by 2025, and employs innovative circular economy principles to reduce waste and extend the lifespan of their products through repair and recycling programs, showcasing their commitment to sustainability in the fashion industry. Timberland's

Table 1 Comparative of sustainability practices

Sustainable practices			
Companies	Animal welfare	Reuse/recycle	Rent
Adolfo Dominguez	✓	✓	✓
Timberland	✓	✓	✓
Anatomie	✗	✗	✗
Love Culture	✗	✗	✗
Duluth Trading Com.	✗	✓	✗
Inditex Group	✓	✗	✗
Mango	✓	✓	✗
Desigual	✗	✓	✗
Cortefiel	✗	✓	✗
H&M	✓	✓	✓
Benetton	✓	✓	✗

Source: Own

animal welfare policy focuses on ensuring the ethical treatment of animals throughout their supply chain.

The company is committed to sourcing materials and products in a responsible manner, avoiding the use of fur and exotic animal skins, and adhering to rigorous standards for the treatment of animals used in leather and wool production. Among the analyzed companies, it is the one that most closely aligns with Adolfo Dominguez's sustainability philosophy.

Its advertising prominently features the concept of reuse as a means of environmental preservation (Fig. 5). Within the scope of the topics currently under discussion, Timberland has instituted an initiative known as "Timberloop." This program is designed to facilitate community engagement in recycling practices by ensuring that the products marketed by the company can either be repurposed for extended utility or, if beyond repair, recycled to generate fresh products. This initiative aims to empower individuals to "close the loop" by maximizing the longevity of products within the supply chain.

Anatomie prioritizes sustainability through several key practices. Overall, their sustainable practices encompass material choices, ethical production, and a long-term view on product durability and environmental responsibility. However, there is no clear evidence that any of the three aspects analyzed (animal welfare, reuse/recycle, rent) constitutes a core axis of their strategy or their communication with consumers.

Love Culture, considered as a fast-fashion retailer, had not been widely recognized for specific sustainable practices in the industry. The company primarily focused on offering affordable, trendy clothing and accessories.

Duluth Trading Company is committed to various sustainable practices in the fashion industry. They emphasize durable and long-lasting clothing designs to reduce the need for frequent replacements, thereby decreasing overall environmental impact. The company sources materials responsibly, often utilizing organic and sustainable fabrics. They also focus on ethical manufacturing, maintaining partnerships with suppliers who adhere to fair labor practices. Additionally, Duluth Trading Company prioritizes efficient shipping and packaging methods to reduce waste and

Source: Timberland web page https://www.timberland.com [18/10/2023]

Fig. 5 Advertising selection related to the concept of reuse. (*Source:* Timberland web page https://www.timberland.com [18/10/2023])

minimize their carbon footprint. These practices collectively reflect their dedication to sustainability and responsible fashion within the industry.

While the *Inditex Group*, the corporate entity overseeing the renowned fashion label Zara, is often associated with fast fashion, its sustainability practices reflect a holistic approach to addressing environmental and social concerns within the fashion industry. The true is that Inditex is garnering recognition for its steadfast dedication to sustainability within the fashion sector. Central to its commitment are several core sustainability practices. Among these practices is the concerted effort to integrate sustainable materials into its product offerings, including organic cotton, Tencel, and recycled fibers. This commitment is underscored by established targets aimed at augmenting the use of these eco-friendly materials, thereby reducing the environmental impact of its products. However, it is also true that Inditex is criticized for contributing to environmental degradation due to its fast-fashion model, which emphasizes frequent clothing turnover. The high production volume, coupled with short product lifecycles, results in increased waste and excessive resource consumption. Critics argue that this approach accelerates textile waste and exacerbates pollution. In fact, regarding recycling or reusing products, Inditex has not a clear program for consumers.

Furthermore, Inditex has aligned itself with the Zero Discharge of Hazardous Chemicals (ZDHC) initiative, demonstrating its dedication to eliminating hazardous substances from the textile and footwear supply chain. In its pursuit of a circular and sustainable fashion model, Inditex has introduced the innovative practice of collecting used clothing from customers in its stores and subsequently recycling these items into new garments. This initiative not only mitigates textile waste but also fosters a more sustainable and eco-conscious fashion ecosystem. However, critics argue that Inditex's rapid fashion turnover encourages overconsumption and a throwaway culture, leading to a disregard for the long-term value and durability of clothing. This cycle can promote a culture of disposability and materialism.

Moreover, Inditex has displayed a commitment to reducing its environmental footprint in its physical stores. This endeavor is achieved through the incorporation of energy-efficient technologies, LED lighting, and the use of sustainable construction materials in store design. These measures collectively contribute to more eco-efficient retail spaces that align with the company's overarching sustainability goals. In addition, Inditex has positioned itself as a water steward by actively working to reduce its water footprint. The implementation of water-saving technologies in its manufacturing processes, combined with the promotion of responsible water management practices throughout its supply chain, reflects the company's proactive approach to addressing water-related environmental concerns. Furthermore, Inditex is engaged in the transition to more sustainable and recyclable packaging materials. This shift is intended to reduce the environmental impact associated with the packaging of its products, a crucial step in minimizing waste and conserving resources.

The company steadfastly shares information concerning its supplier factories, their environmental practices, and labor conditions. But, although Inditex has made efforts to improve supply chain transparency, critics argue that the company should do more to ensure that its supplier factories meet high ethical and sustainability standards. Inditex is unwavering in its dedication to ensuring fair labor practices within its supply chain. This includes the provision of fair wages, the maintenance of safe working conditions, and the empowerment of workers. These endeavors epitomize the company's commitment to social responsibility and ethical labor practices. However, some concerns have been raised about the wage disparities within the fashion industry, with Inditex being no exception. Critics point to the high salaries of executives in contrast to the low wages of factory workers, highlighting income inequality within the company. The company has also sharpened its focus on achieving eco-efficient logistics, a strategy aimed at optimizing transportation and supply chain operations to reduce carbon emissions. This initiative encompasses the use of more sustainable transportation methods, marking a significant step toward mitigating the environmental impact of its distribution network.

Lastly, Inditex includes in its policy ethical standards in the use of animal products in accordance with the "Five Freedoms" for the ethical and conscientious treatment of animals, widely adopted and recognized as a standard for animal welfare globally.

Mango, a Spanish fashion company, has implemented a range of sustainability practices. The company emphasizes sustainable materials, including organic and recycled fabrics, and has committed to using 100% sustainable cotton by 2025. Mango is actively reducing their environmental impact by implementing energy-efficient measures in their stores and warehouses. They also focus on ethical and fair labor practices, collaborating with suppliers who adhere to strict labor standards. The company has introduced recycling programs and a "Take Action" initiative that promotes responsible consumption and clothing care. Mango's comprehensive sustainability efforts aim to address environmental and social concerns, making them a more responsible player in the fashion industry. In fact, Mango is running advertising campaigns with the slogan "recycle your clothes today for a better tomorrow."

In addition, Mango is embarking on a clothing rental pilot program hosted on its website under the banner of Mango Renting (www.mangorenting.com). This service will encompass approximately 40 clothing options, subject to monthly updates, with the objective of enhancing versatility and flexibility for Mango's clientele in the selection of ensembles suitable for their various occasions. In the same line, Mango StartUp Studio is investing in the party clothing rental start-up La Más Mona (i.e., a colloquial expression in Spanish which means the cutest girl or the most adorable girl).

Desigual, another Spanish fashion company, has started to incorporate sustainability practices into its operations, with a focus on the following key areas: The firm is working to reduce the environmental footprint by using sustainable materials like organic cotton and recycled fibers, as well as implementing responsible production methods to minimize water and energy consumption. Desigual is also emphasizing ethical sourcing and fair labor practices within their supply chain. Additionally, Desigual is making efforts to reduce waste and encourage responsible consumption having launched initiatives like clothing recycling programs.

Cortefiel, one more Spanish fashion company, actively pursues a range of sustainability initiatives. It places a significant emphasis on responsibly sourcing materials, particularly sustainable and organic fabrics, while also advocating for eco-friendly production methods. The company is dedicated to reducing its carbon footprint through the implementation of energy-efficient measures within its retail outlets and distribution centers. Ethical practices are at the forefront of their supply chain, including the promotion of fair labor conditions and labor rights. Their goals include waste reduction and the promotion of responsible consumption, exemplified by the introduction of clothing recycling programs.

H&M has implemented several sustainable practices related to animal welfare, reuse, recycling, and rental services. In terms of animal welfare, the company has a commitment to using sustainable and responsibly sourced materials, including a goal to use 100% sustainably sourced or recycled materials by 2030, reducing their reliance on animal-derived materials. Regarding reuse and recycling, H&M has established in-store clothing collection programs, enabling customers to return old garments for recycling or reuse. The company also promotes a circular fashion approach through their "Conscious Collection" and "Close the Loop" initiatives, which feature products made from recycled materials. Additionally, H&M has ventured into clothing rental services, offering customers the option to rent clothing items, thereby extending the lifecycle of the products and reducing fashion waste. These practices collectively reflect H&M's commitment to more sustainable and ethical fashion.

Benetton, an Italian fashion company, had not been prominently recognized for specific sustainable practices related to animal welfare, reuse, recycling, or rental services in the public domain. However, the company has implemented several sustainable practices. It is focused on responsible sourcing of materials, with efforts to use sustainable and organic fabrics, reducing their environmental impact. The company has a range of products called Green B produced with recycled materials. Benetton also has prioritized fair labor practices and ethical treatment within their supply chain. The company has aimed to minimize waste through recycling initiatives and has introduced clothing collection programs in its stores. Furthermore, Benetton promoted responsible consumption by encouraging customers to buy durable and timeless pieces.

6 Conclusion

The textile industry has encountered enduring challenges pertaining to social and environmental sustainability throughout its historical trajectory. In response to these challenges, numerous textile enterprises are increasingly integrating CSR initiatives into their operational frameworks. This integration is not solely driven by moral imperatives but is also strategically motivated to align with dynamic consumer expectations and fortify brand reputation. The convergence of slow-fashion principles with CSR assumes particular significance in addressing these challenges, underscoring the pivotal role of sustainable and ethical practices in the textile industry's processes of production and consumption. In fact, the fashion industry has been undergoing a significant transformation in recent years, driven by a growing awareness of environmental and social issues. Several sustainable trends have emerged, reflecting a shift toward more responsible and eco-conscious practices in the fashion world.

In this chapter, we have introduced slow fashion as a sustainable alternative to fast fashion. We have presented the case of the Spanish company Adolfo Dominguez as an example of a fashion industry company highly committed to sustainability. The described advertising campaigns illustrate the company's concern for animal welfare, garment recycling and reuse, responsible consumption, and the awareness that sometimes less is more.

We believe that the company's recent initiatives, particularly those related to garment rental, signify another step in its progression toward genuine responsible consumption. We highlight that the company would be making a shift from production to service. By pioneering the social innovation inherent in these initiatives, we believe that Adolfo Dominguez is at the forefront of the global responsible fashion sector.

In the comparative study among companies in the sector, we have observed that it is a fact that many fashion brands have been increasingly incorporating sustainability concerns and recycling/reuse themes into their strategies. Some of them are also incorporating sustainable messages into their advertising and messaging in recent years as consumer demand for eco-friendly practices and products has grown.

We believe that Adolfo Dominguez is a leading company that can guide the steps of others toward a more environmentally conscious and respectful market, contributing to dignify the textile industry. To finish, we can highlight several sustainable trends for the fashion industry. One of the most prominent is the shift from a linear to a circular fashion economy. This approach emphasizes recycling, upcycling, and reusing materials to minimize waste. In addition, the secondhand and vintage mentality, thrifting, and the resale market are gaining popularity. Companies and brands like Adolfo Dominguez are increasingly designing products with longevity in mind, promoting repair, and take-back programs to keep garments in circulation.

The use of sustainable and eco-friendly materials is also gaining traction. This includes fabrics made from organic cotton, recycled polyester, and other renewable resources. Innovations such as lab-grown leather and plant-based alternatives are starting to reduce the environmental impact of fashion. There's a push for greater transparency within the industry, and certification standards like fair trade and certifications for organic materials are helping to validate these claims.

But without doubts, the slow-fashion movement is promoting sustainability, focusing on quality over quantity, encouraging consumers to buy fewer, higher-quality items that last longer. An essential complement to this is the consumer education. We believe that an informed and engaged consumer base is a crucial driver of sustainability in fashion. Increasingly, consumers are educating themselves about the impact of their choices and demanding more sustainable options. But also, responsible advertising and the messages of companies in their advertising contribute to this awareness. This approach, with Adolfo Dominguez as leader, contrasts with fast fashion, which relies on rapid turnover of inexpensive, disposable garments and an unsustainable future for everyone.

References

Abbate, S., Centobelli, P., Cerchione, R. C., Nadeem, S. P., & Riccio, E. (2023). Sustainability trends and gaps in the textile, apparel and fashion industries. *Environment, Development and Sustainability*, 1–23. https://doi.org/10.1007/s10668-022-02887-2

Adamkiewicz, J., Kochańska, E., Adamkiewicz, I., & Łukasik, R. M. (2022). Greenwashing and sustainable fashion industry. *Current Opinion in Green and Sustainable Chemistry, 38*, 100710.

Barnes, L., & Lea-Greenwood, G. (2006). Fast fashioning the supply chain: Shaping the research agenda. *Journal of Fashion Marketing and Management: An International Journal, 10*(3), 259–271.

Caro, F., & Martínez-de-Albéniz, V. (2015). Fast fashion: Business model overview and research opportunities. In N. Agrawal & S. Smith (Eds.), *Retail supply chain management* (International series in operations research & management science) (Vol. 223). Springer. https://doi.org/10.1007/978-1-4899-7562-1_9

Galan-Ladero, M. M., & Sánchez-Hernández, M. I. (2022). Corporate donation behavior during the covid-19 pandemic. A case-study approach in the multinational inditex. *Heliyon, 8*(12), e12188.

Grazzini, L., Acuti, D., & Aiello, G. (2021). Solving the puzzle of sustainable fashion consumption: The role of consumers' implicit attitudes and perceived warmth. *Journal of Cleaner Production, 287*, 125579.

Joy, A., Sherry, J. F., Jr., Venkatesh, A., Wang, J., & Chan, R. (2012). Fast fashion, sustainability, and the ethical appeal of luxury brands. *Fashion Theory, 16*(3), 273–295.

Jung, S., & Jin, B. (2016). From quantity to quality: Understanding slow fashion consumers for sustainability and consumer education. *International Journal of Consumer Studies, 40*(4), 410–421.

Lee, J. E., Hur, S., & Watkins, B. (2018). Visual communication of luxury fashion brands on social media: Effects of visual complexity and brand familiarity. *Journal of Brand Management, 25*, 449–462.

Liu, S. Y., Napier, E., Runfola, A., & Cavusgil, S. T. (2020). MNE-NGO partnerships for sustainability and social responsibility in the global fast-fashion industry: A loose-coupling perspective. *International Business Review, 29*(5), 101736.

Niinimäki, K., Peters, G., Dahlbo, H., Perry, P., Rissanen, T., & Gwilt, A. (2020). The environmental price of fast fashion. *Nature Reviews Earth & Environment, 1*(4), 189–200.

Peters, G., Li, M., & Lenzen, M. (2021). The need to decelerate fast fashion in a hot climate-A global sustainability perspective on the garment industry. *Journal of Cleaner Production, 295*, 126390.

Quiles-Soler, C., Martínez-Sala, A. M., & Monserrat-Gauchi, J. (2023). Fashion industry's environmental policy: Social media and corporate website as vehicles for communicating corporate social responsibility. *Corporate Social Responsibility and Environmental Management, 30*(1), 180–191.

Štefko, R., & Steffek, V. (2018). Key issues in slow fashion: Current challenges and future perspectives. *Sustainability, 10*(7), 2270.

Todeschini, B. V., Cortimiglia, M. N., Callegaro-de-Menezes, D., & Ghezzi, A. (2017). Innovative and sustainable business models in the fashion industry: Entrepreneurial drivers, opportunities, and challenges. *Business Horizons, 60*(6), 759–770.

Vänskä, A., & Gurova, O. (2022). The fashion scandal: Social media, identity and the globalization of fashion in the twenty-first century. *International Journal of Fashion Studies, 9*(1), 5–27.

Recycling: An Approach Towards Sustainability and to Maintain Corporate Social Responsibility

Sankar Roy Maulik and Antu Chandra

Abstract Today, environmental protection has become a global challenge. Textile and fashion industries are also a part of this challenge since they use many dyes, chemicals and other accessories for their entire supply chain. The world is becoming increasingly concerned about the planet's health due to the alarming use of those and the increased effluent load. Corporate social responsibility is a business strategy that encourages businesses to work for the betterment of mankind and the environment rather than degrading them. It has become increasingly significant as consumers, investors and society at large expect organisations to go beyond profitability and consider their impact on the environment and society.

It can enhance several societal facets and advance a favourable brand image for businesses. Sustainable clothing materials can be produced from eco-friendly resources such as sustainably cultivated fibre crops or recycled materials. Purchasing an outfit affects the environment, more than a fashion statement. Thus, when it comes to apparel and textiles, the priority of the consumer is quality over quantity, by selecting eco-friendly fabrics and sustainable textiles.

To create sustainable textiles, the initiative is taken to use organic cotton and recycled polyester yarn, and fabrics are produced on a handloom. The colour scheme used to create a line of formal office clothing and accessories is based on the colour forecast for 2023–2024.

Keywords Corporate social responsibility · Handloom · Organic · Polyester · Recycled · Sustainable · Textiles

S. Roy Maulik (✉) · A. Chandra
Department of Silpa-Sadana, Visva-Bharati (A Central University), Sriniketan, India
e-mail: sankar.raymaulik@visva-bharati.ac.in

S. S. Muthu (ed.), *Corporate Social Responsibility in Textiles and Fashion*,
Sustainable Textiles: Production, Processing, Manufacturing & Chemistry,
https://doi.org/10.1007/978-3-031-61099-8_7

1 Introduction

The process by which an organisation balances its social, environmental and financial obligations while satisfying the needs of shareholders and other stakeholders is known as corporate social responsibility (CSR). Corporate social responsibility is used to describe an organisation's efforts to improve society and the environment. It is becoming an increasingly important part of making fashion and textiles more sustainable. The organisation and stakeholders have been pressured to adopt strategies in their business, which are sustainable since only economic advantage alone cannot be beneficial in the long run. Corporate social responsibility (CSR) is one of the long-term sustainable strategies that every organisation must apply. CSR gained popularity due to its ability to embrace all aspects of sustainability, whereas other sustainable strategies may have limitations or fail (Pagell et al., 2008; Ageron et al., 2012). CSR is relevant to an array of businesses throughout the globe, including textiles and fashion. These sectors have recently experienced a major crisis due to CSR issues (Shen et al., 2015). Fashion brands put a tremendous effort into promoting their brand image through corporate social responsibility practices. They are using various media platforms and communication techniques to promote their CSR campaign.

Consumers would boycott a company if it participated in unethical business practices, and most global consumers expect companies to act responsibly to address a wide range of social and environmental challenges. Corporate social responsibility can be categorised as environmental responsibility, ethical or human rights responsibility, philanthropic responsibility and economic responsibility.

Presently, textile producers and fashion brands are more conscious of their societal and environmental responsibilities, and they are implementing successful and innovative eco-friendly corporate social responsibility initiatives to reduce environmental impact and thus promote textile sustainability, which indirectly has a positive impact on the growth of that organisation or brand. Corporate social responsibility activities are key in making clothing brands sustainable and help improve society and the environment. These sustainable initiatives should be addressed throughout the textile value chain. CSR is also beneficial in spreading awareness about the green side of a brand that ultimately results in significant benefits in terms of financial as well as societal. These benefits include increased brand recognition, a stronger global corporate reputation and greater opportunities to raise public awareness about global environmental concerns. A well-designed corporate social responsibility plan implemented in the textile industry will result in the more effective use of natural resources, more sustainable growth and higher economic performance. Apart from these primary benefits, the organisation can gain from the marketing opportunities and goodwill created by environmentally friendly and environmentally conscious business practices. Nowadays, the fashion brands and textile industries are very much concerned towards their responsibilities to society. The current damage due to fast fashion, excessive use and production of textiles by consumers and producers play a significant role in maintaining proper CSR. Textile industries are presently investing in areas that include the reduction of carbon footprint, dealing with fair trade items, participating in social causes, proper labour policies and charitable projects. Successful CSR programs have an impact on the entire textile supply chain. Positive outcomes of CSR benefit public perception, consumer trust and enhanced brand awareness. Corporate social responsibility is no longer an option; it is a must for a professional modern textile business.

Sustainability has become one of the most popular buzzwords in recent decades. It can be described as a 'boundary' term, in which science and politics are interwoven. Sustainability and sustainable development have produced an intellectual debate among academia and politicians since the United Nations Conference on Environment and Development (UNCED), held in Rio in 1992 (Scoones, 2007). Sustainable development is a development that meets the needs of the present without compromising the ability of future generations to meet their own needs (Scoones, 2007). The concept of sustainability has been accepted as multidimensional; its numerous components have brought to light various discourses over time and are frequently discussed independently. This separation has, in some circumstances, reduced the real implementation of sustainability to mere rhetoric (Giovannoni & Fabietti, 2013). The primary goal of sustainability is to maximise utility by extending the life of products. Technical and economic requirements, as well as the necessity for waste prevention and recycling, are crucial in terms of sustainability. In the present study, dyed recycled polyester yarns and organic cotton yarns were used to produce value-added handloom products. *Terminalia chebula* (myrobalan) and *Indigofera tinctoria* L. were used as natural colourants for dyeing organic cotton yarn in the presence and absence of different innocuous inorganic salts or mordants. Using recycled polyester yarn on handloom involves blending recycled materials in traditional techniques. The use of recycled yarn enhances the product's durability. The present work also highlights the experimentation with weaving patterns and designs to showcase the unique texture of handloom fabrics while promoting sustainability. This idea helps collaborate with skilled artisans to balance traditional craftsmanship and modern eco-friendly materials.

2 Recycling and Recycled Polyester

The shift towards eco-friendly and sustainable materials is evident, with brands adopting organic cotton, recycled fibres and innovative textiles produced with reduced environmental impact. Circular economy principles have gained momentum, promoting reuse, recycling and waste reduction. Brands are designing products for longevity, offering repair services and implementing take-back programs.

Recycling helps promote a circular economy by establishing a closed-loop production cycle. It can be broadly classified into (i) primary, (ii) secondary, (iii) tertiary and (iv) quaternary approaches. Primary recycling, also known as closed-loop recycling, is the recycling of a product to its original state and is considered as the most beneficial recycling method. Secondary recycling, also known as the open-loop process, on the contrary, is the recycling of waste into a new product of different application areas with lower physical, mechanical and chemical properties that may have a different application compared to the original. The life cycle of the products is extended, and disposal processes are delayed with open-loop recycling. Materials which can be used in the open-loop recycling method for the textile and garment industry are pre-consumer textile waste, post-consumer textile waste and post-consumer waste polyethylene terephthalate (PET) bottles. Tertiary recycling is the process of converting waste into simple chemicals or fuels through pyrolysis, gasification or hydrolysis. Quaternary recycling is a process of incineration.

The rapid growth in application due to high tensile and impact strength, chemical resistance, thermal stability and colouring ability makes virgin polyester one of the most important engineering polymers nowadays. It is commonly used for textiles and is regarded as an excellent material for various purposes including liquid containers (Awaja & Pavel, 2005). Terylene, the first polyester fibre, was produced shortly after the discovery of nylon and was termed a 'miracle fibre' due to its ease of laundering and wrinkle-resistance properties. The growing market demands for polyester forced various organisations to produce their brand and presently accounts for approximately half of the fibre market globally (Sustainable Materials Guide //04 Polyester, 2023). The synthetic polymer-based fibres do not require agricultural land, but they have significant environmental consequences in terms of an enormous amount of energy required for heating and cooling, which substantially contributes to greenhouse gas emissions. Polyester emits nearly three times the amount of greenhouse gases as cotton and is also non-biodegradable; as a result, the majority of polyester clothing ends up in landfills after use. Polyester also faces the issue of shredding microfibres and small particles with every wash. These microplastics are found in every ecosystem around the world, and it is estimated that around half a million tons of microplastics are discharged into the oceans yearly through wastewater, and 35% of primary microplastics are released during washing cycles. These chemicals and plastic particles are then eaten by small fish and consequently moved into larger food systems, including human food cycles (Sustainable Materials Guide //04 Polyester, 2023).

The fibre and selection of raw materials in the textile supply chain across its lifecycle account for a significant impact on the environment. Appropriate selection of fibres is under the control of fashion brands and therefore represents a key point for a sustainable alternative. Thus, the use of recycled fibre, yarn and fabric is the key element for designing new products. Recycled materials used in textile and apparel products can be obtained throughout the textile and apparel supply chain and post-consumer collection methods (Leonas, 2017).

Environmental pressure to improve waste management resulted in the growth of recycling industries. The slow rate of natural decomposition and the unavailability of known microorganisms for the faster decomposition of PET are the main driving forces for producing recycled polyester.

Recycled polyester is commonly produced either mechanically or chemically using pre- or post-consumer waste. However, according to the Textile Exchange Preferred Fibre Materials Report 2017, only 7% of all polyester used is recycled.

Recycled polyester produced from PET bottles performs equally well as 100% virgin polyester. Recycling PET bottles to create the fabric prevents it from going to landfill, and it causes less air, water and soil contamination compared to using non-recycled polyester. Waste bottles are collected, cleaned and sorted by colour in the mechanical recycling method. The labels are removed, bottles are converted to PET flakes, and, finally, PET flakes are melted, extruded and spun into a polyester filament or staple fibre known as recycled polyester or r-PET. In the case of chemically recycled, the PET flakes are broken down into the parent chemical molecule, and the process of the original polyester manufacturing system is followed. It is estimated that 19 PET bottles would be enough to produce fabric for one extra-large T-shirt, indicating the significant amounts of bottles that can be saved from being sent to landfill or incineration. Polyester fabric recycling is less common, and here the polyester scraps and textiles are broken down into fibre components that can then be spun into recycled yarns. Designing clothes with recycled polyester has several advantages. Polyester accounts for more than half of the fibres used in the textile and garment industries, and partial replacement always helps reduce dependency on fossil fuels, thus helping brands reduce greenhouse gas emissions throughout the supply chain. r-PET also contributes to the reduction of solid waste generated in the fashion sector. Another benefit is that recycled polyester has a smaller carbon footprint than virgin polyester. So, using recycled polyester may help brands meet their climate targets and contribute to Sustainable Development Goals.

The use of plastic to convert into ultimate fabric will undoubtedly minimise the dumping of plastics in landfills and oceans. This will also minimise greenhouse gas emissions and dependency on fossil fuels. The use of recycled plastics will certainly change the patterns in fashion industries due to its known advantages, viz. less energy and water consumption, minimising air pollution and dumping plastic bottles in landfills. Some of the giant brands in fashion, viz. GAP, H&M, Adidas, etc., increased their use of r-PET in their product line (Telli et al., 2012). Researchers also found no significant differences in the properties of the garments developed using cotton/r-PET and cotton/virgin PET of the same counts (Necef et al., 2013). The properties of yarns obtained from r-PET fibres and blends obtained by recycling PET bottle wastes have also been reported in the literature. Tensile strength, elongation, evenness, hairiness properties and imperfection index values of yarns spun using nine different blend ratios were measured, and their results were evaluated (Telli & Özdil, 2013). The latest technologies, viz. dissolution and depolymerisation techniques of recycling methods used in the production of PET and nylon fibres, are also reported in the literature (Lu & Hamouda, 2014).

Researchers also compared the properties of recycled PET against virgin PET fibres. The results show that the surface morphology of both PET fibres is similar, while recycled PET fibres have a higher tensile strength and greater elongation at break (He et al., 2015). Figure 1 shows a schematic representation of producing recycled polyester yarn from post-consumer waste, i.e. polyester bottles (Sustainable Materials Guide //04 Polyester, 2023).

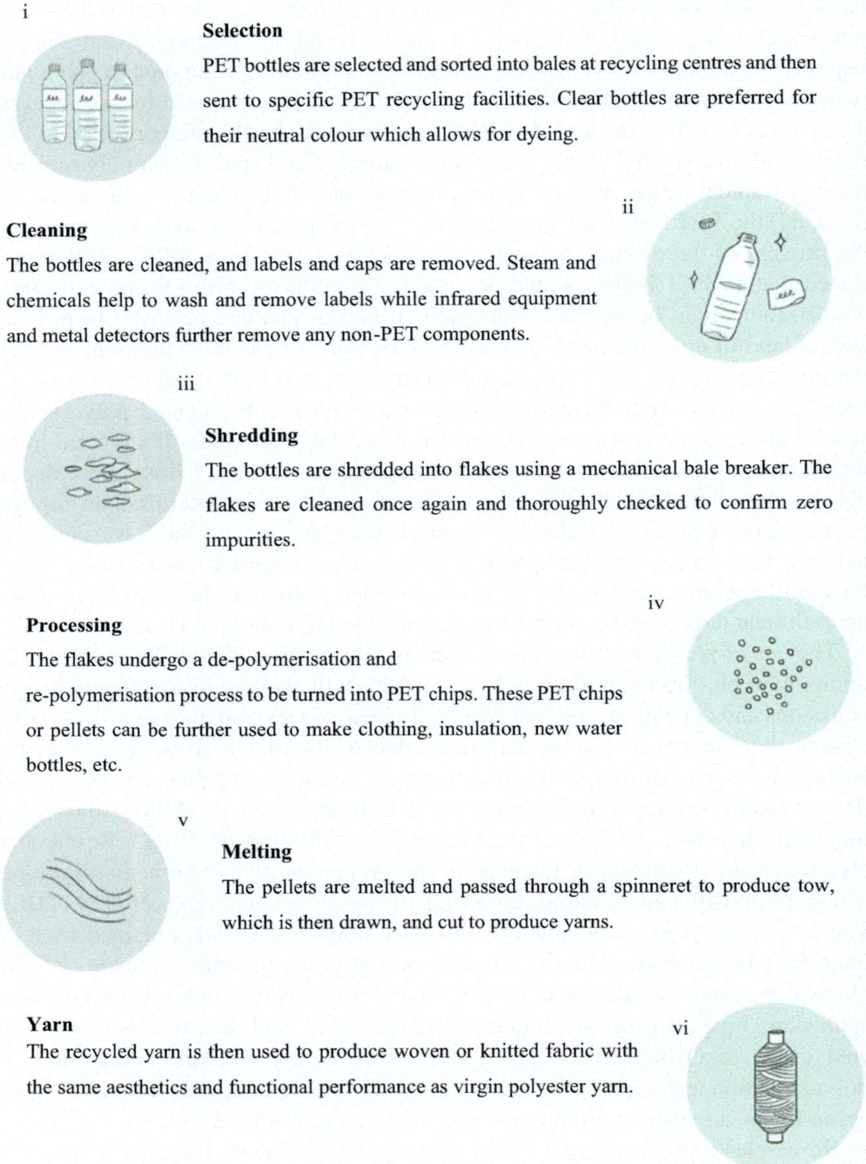

Fig. 1 Schematic representation of producing r-PET from post-consumer waste

3 Organic Cotton

Organic cotton by definition refers to any cotton that is cultivated and certified according to the organic agriculture standards. It includes cotton that is grown, harvested and processed without the use of chemical fertiliser, pesticide or herbicide, growth regulator or defoliant (Mageshwaran et al., 2019).

Organic cotton has provided significant price premiums for growers willing to meet the many challenges inherent in its production without the aid of conventional pesticides and commercial fertilisers (Mageshwaran et al., 2019; Guerena & Sullivan, 2003). World production of organic cotton amounts is about 0.03% of global cotton production. The major organic cotton-producing countries are India, China, Kyrgyzstan, Turkey and the USA (Mageshwaran et al., 2019). Europe and the USA contribute to the maximum use of organic cotton (Peter, 2002). It was initially planted to provide sustainable, ecological and biodynamic agriculture. Organic cotton fosters and enhances biodiversity and biological cycles, benefiting human health and the environment. As a result, even though the qualities of organic cotton fibre are inferior to those of ordinary cotton fibre, global production of organic cotton is fast increasing.

4 The Objective of This Study

Clothing is one of the most basic human necessities, and the global production of textile fibres and clothing fulfils the need every day. The choice for apparel has evolved over time, and nowadays consumers prefer to use apparel that is attractive and inexpensive. Sustainable items are becoming increasingly popular, and consumers want ‘green products’. In this context, the handloom products are becoming increasingly popular among various brands and designers. The present work is focused on design development for apparel and accessories using recycled polyester and organic cotton yarn in handloom. Recycled polyester offers the opportunity to reduce energy and waste implications for organisations trying to switch to more sustainable types of textiles. This article examines the significance of successful, environmentally friendly corporate social responsibility initiatives that help both businesses and the environment.

5 Methodology

5.1 Materials

In this study, recycled PET (r-PET) and organic cotton yarns of the following specifications as mentioned in Table 1 were used.

Table 1 Specifications of yarns used in this study

Types of yarn		Yarn count (Ne)
Recycled polyester	Dark oak	$2/20^s$
	Jade	$2/20^s$
	Lazuli blue	$2/20^s$
Organic cotton		$2/20^s$

5.2 *Dyes and Chemicals*

Terminalia chebula (myrobalan) and *Indigofera tinctoria* L. (indigo) were used as natural dyes for dyeing organic cotton yarn. Ferrous sulphate was used as a mordant during the application of natural dye, i.e. myrobalan. Sodium carbonate, sodium hydroxide, sodium hydrosulphite of LR grade and non-ionic detergent of commercial grade were also used during this study.

5.3 *Accessories*

Polyester and cotton fabric, chain, metal zipper, leather and wooden buttons were also used during the work as and when required.

5.4 *Methods*

5.4.1 Scouring of Yarn

Organic cotton yarn was treated with sodium carbonate (3%) and detergent (0.5%) at 90°C for 30 min to improve absorbency. The yarn was thoroughly washed with normal water and dried at room temperature.

5.4.2 Dyeing with Natural Dyes

The raw materials for dyeing were collected from local resources. Myrobalan was dried in the absence of direct sunlight and crushed with the help of a pulveriser. It was then boiled in a specific volume of water at 90°C for 30 min. The scoured organic cotton yarn was treated with dye extract at 60°C for 15 min, thus mordanting with ferrous sulphate at room temperature for 5 min. After dyeing, the yarn was washed with non-ionic detergent, washed in normal water and finally dried at room temperature.

Dyeing of organic cotton was carried out in a vat-containing indigo, sodium hydrosulphite and sodium hydroxide. The vatting and solubilising process was performed at room temperature for 10 min. Organic cotton yarns were carefully dipped in this solution for 20 min, followed by oxidation in the solution containing hydrogen peroxide and acetic acid for another 10 min. Soaping of the dyed yarn was done employing non-ionic detergent for 5 min at 50°C (Roy Maulik, 2021).

5.4.3 Handloom Weaving

The handloom has been the pride of India's traditional and cultural heritage. Handloom weaving was performed on frame looms and hobby looms following standard preparatory to weaving processes, viz. bobbin winding, beam warping, drawing and denting, looming, pirn winding, etc., before weaving (Roy Maulik et al., 2016). In all cases, organic cotton and/or recycled polyester were used as warp yarn, whereas recycled polyester yarns were used as weft. The major processes involved in the making of hand-woven cloth involve (i) dyeing of yarns, (ii) bobbin winding, (iii) drumming, (iv) warping, (v) pirn winding and (vi) weaving. The fabrics were produced according to the design and the various stages of the fabric production process as described in Fig. 2.

5.4.4 Fabric Cutting and Stitching

Fabric cutting was done as per pattern to make garments and accessories. Those fabrics were stitched in a single needle lock stitch machine to produce garments and other accessories.

5.4.5 Samples Details

Sample No. 1

- Weave: Broken twill

Material: Warp—recycled polyester
Weft—recycled polyester
Reed count: 48 s stockport
Denting: Selvedge—4/dent
Body—1/dent
Drafting: 1,2,3,4,4,3,2,1
Lifting: (1,2,3,4,3,2,1)
Yarns count: $2/20^s$ (warp and weft)
Areal density: 216 g/m^2

2(a)

2(b)

2(c)

2(d)

2(e)

2(f)

2(g)

Fig. 2 The entire working process of producing clothes and accessories. (**a**) Dyeing of yarns, (**b**) Bobbin winding, (**c**) Warping, (**d**) Weaving on handloom, (**e**) Woven fabric, (**f**) Fabric cutting, (**g**) Stitching

Sample No. 2

- Weave: Plain

Material: Warp—organic cotton
Weft—recycled polyester
Reed count: 48 s stockport
Denting: Selvedge—4/dent
Body—2/dent
Drafting: 1,2,3,4
Lifting: (1,3), (2,4).
Yarns count: $2/20^s$ (warp and weft)
Areal density: 174 g/m^2

Sample No. 3

- Weave: Twill

Material: Warp—organic cotton
Weft—recycled polyester
Reed count: 48 s stockport
Denting: Selvedge—4/dent
Body—1/dent
Drafting: 1,2,3,4
Lifting: (1,2,3,4)
(1,3), (2,4)

Yarns count: $2/20^s$ (warp and weft)
Areal density: 212 g/m^2

Sample No. 4

- Weave: Twill

Material: Warp—organic cotton
Weft—organic cotton and recycled polyester
Reed count: 48 s stockport
Denting: Selvedge—4/dent
Body—1/dent
Drafting: 1,2,3,4
Lifting: (1,2,3,4)
(1,3), (2,4)
Yarns count: $2/20^s$ (warp and weft)
Areal density: 196 g/m^2

5.4.6 Mood Board

The mood board is a prevalent medium to attract a person. It is a collage consisting of images, text and sample objects in a composition. Mood boards can be based on a set topic or can be any material chosen on a random basis (Roy Maulik et al., 2019). Figure 3 depicts a mood board for the present collection.

Illustration

Fig. 3 Mood board

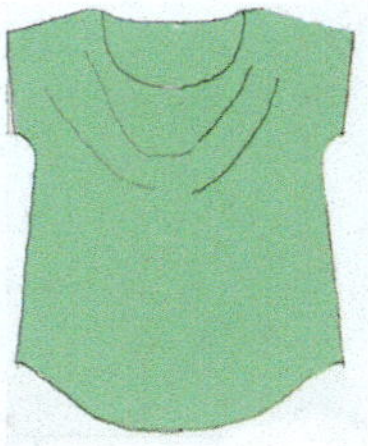

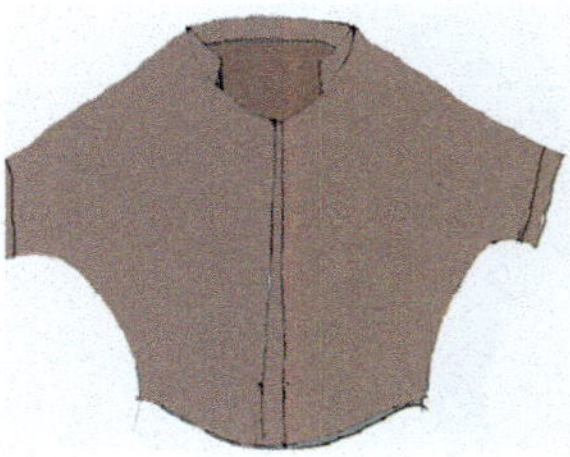

5.4.7 Cost Sheet and Product Details

Style no.	Weave	Yarn	Colour	
1 (blazer)	Plain	**Warp**	**Weft**	Light blue
		Organic cotton	Recycled polyester	

Raw materials cost (Rs.) (yarn + fabric making)	400
Other materials cost (Rs.)	150

Stitching cost (Rs.)	600
Total (Rs.)	1150
Selling price (20% of the total price) (Rs.)	1380

Blazer

Style no.	Weave	Yarn		Colour
2 (jacket)	Broken twill	**Warp**	**Weft**	Green
		Recycled polyester	Recycled polyester	

Raw materials cost (Rs.) (yarn + fabric making)	400
Other materials cost (Rs.)	150
Stitching cost (Rs.)	500
Total (Rs.)	1050
Selling price (20% of the total price) (Rs.)	1260

Jacket

Style no.	Weave	Yarn		Colour
3 (formal shirt)	Plain	**Warp**	**Weft**	Light green
		Organic cotton	Recycled polyester	

Raw materials cost (Rs.) (yarn + fabric making)	600
Other materials cost (Rs.)	50
Stitching cost (Rs.)	200
Total (Rs.)	850
Selling price (20% of the total price) (Rs.)	1020

Formal Shirt

Style no.	Weave	Yarn		Colour
4 (ladies top)	Plain	**Warp**	**Weft**	Light green
		Organic cotton	Recycled polyester	

Raw materials cost (Rs.) (yarn + fabric making)	300
Others materials cost	–
Stitching cost (Rs.)	200
Total (Rs.)	500
Selling price (20% of the total price) (Rs.)	600

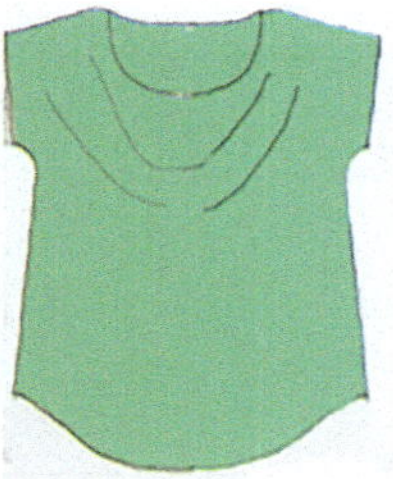

Ladies Top

Style no.	Weave	Yarn		Colour
5 (ladies top)	Plain	**Warp**	**Weft**	Light maroon
		Organic cotton	Recycled polyester	

Raw materials cost (Rs.) (yarn + fabric making)	300
Other materials cost (Rs.)	50
Stitching cost (Rs.)	200
Total (Rs.)	550
Selling price (20% of the total price) (Rs.)	660

Ladies Top

Style no.	Weave	Yarn		Colour
6 (stole)	Twill	**Warp**	**Weft**	Grey, maroon
		Organic cotton	Recycled polyester	

Raw materials cost (Rs.) (yarn + fabric making)	600
Others materials cost	–

Stitching cost	–
Total (Rs.)	600
Selling price (20% of the total price) (Rs.)	720

Stole

Style no.	Weave	Yarn		Colour
7 (stole)	Twill, plain	**Warp**	**Weft**	Blue and yellow
		Organic cotton	Organic cotton	

Raw materials cost (Rs.) (yarn + fabric making)	700
Others materials cost	–
Stitching cost	–
Total (Rs.)	700
Selling price (20% of the total price) (Rs.)	840

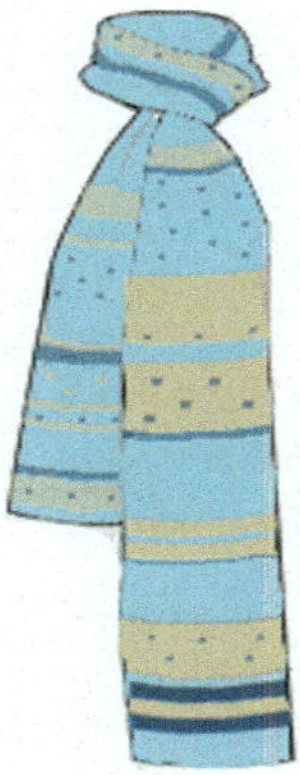

Stole

Look Book

6 Conclusion

Corporate social responsibility (CSR) is a concept in which organisations combine environmental and social issues to maintain a balance of economic, environmental and social commitments while serving the needs of shareholders and stakeholders. It is playing a vital role in making fashion more sustainable. Due to the pressure from consumers, fashion brands are very much involved in promoting their corporate social responsibility campaigns through various media platforms. Brands are also tirelessly involved in promoting their images. Today, producers and brands in the textile sector are more aware of their societal and environmental responsibilities and are implementing successful and innovative corporate social responsibility initiatives. Successful eco-friendly corporate social responsibility policies can help reduce environmental impact and increase the sustainability of textiles while also benefiting the business.

Corporate social responsibility (CSR) and sustainability have undergone a transformative evolution in the textile and fashion industry, and the industry now stands at a critical juncture.

These industries have recognised the urgency of adopting sustainable practices, driven by consumer awareness, changing regulations and the imperative to address global issues such as climate change and labour rights. Brands are increasingly integrating CSR into their core strategies, emphasising ethical sourcing, responsible production and transparent supply chains.

Ensuring fair wages, safe working conditions and respect for human rights is integral to CSR efforts. Brands are emphasising supplier audits, worker empowerment and transparency in the supply chain. They are actively reducing their carbon footprint, water usage and chemical pollutants through responsible production processes and certifications like GOTS (Global Organic Textile Standard) and OEKO-TEX. Brands are also engaging consumers in responsible consumption, promoting sustainable fashion choices and providing information about the environmental and social impact of their products.

The present research is an attempt to make textiles more sustainable. The primary goal of the project, which was to develop garments and accessories out of recycled yarn, was also met. The entire line was handcrafted and made using organic cotton and recycled polyester yarn. In the work, handloom, one of the most essential methods of making sustainable textiles, was used. The demand for sustainable textiles has risen significantly, and consumers' choices and needs are influenced by their comfort and accessibility. Thus, many weave patterns have been developed through this research, and the items are made utilising a 100% sustainable approach with a significant niche market. Corporate social responsibility is no longer an option for a professional modern textile organisation; it is a need.

The growing importance of CSR and sustainability in the textile and fashion industry emphasises the need for a fundamental shift in business practices. The present work not only contributes to a more responsible and ethical industry but also meets the demands of a conscious and informed consumer base, shaping a more sustainable future for fashion and textiles. In conclusion, corporate social responsibility in textile and fashion brands goes beyond profit and aims to create a positive impact on society and the environment. The present work may successfully integrate CSR into a business model not only enhancing their reputations but also contributing to a more sustainable and ethical industry.

References

Ageron, B., Gunasekaran, A., & Spalanzani, A. (2012). Sustainable supply management: An empirical study. *International Journal of Production Economics, 140*, 168–182.

Awaja, F., & Pavel, D. (2005). Recycling of PET. *European Polymer Journal, 41*(7), 1453–1477.

Giovannoni, E., & Fabietti, G. (2013). What is sustainability? A review of the concept and its applications. In C. Busco, M. Frigo, A. Riccaboni, & P. Quattrone (Eds.), *Integrated reporting*. Springer. https://doi.org/10.1007/978-3-319-02168-3_2

Guerena, M., & Sullivan, P. (2003). *Organic cotton production* (P. Williams & D. Zodrow, Ed., pp. 1–24). Retrieved October 26, 2023, from https://attra.ncat.org/wp-content/uploads/2019/05/cotton.pdf

He, S., We, M., Lu, M., et al. (2015). Characterization of virgin and recycled poly (ethylene terephthalate) (PET) fibres. *The Journal of the Textile Institute, 106*(8), 800–806.

Leonas, K. K. (2017). The use of recycled fibres in fashion and home products. In S. Muthu (Ed.), *Textiles and clothing sustainability. Textile Science and Clothing Technology*. Springer. https://doi.org/10.1007/978-981-10-2146-6_2

Lu, J. J., & Hamouda, H. (2014). Current status of fibre waste recycling and its future. *Advances in Materials Research, 878*, 122–131.

Mageshwaran, V., Satankar, V., Shukla, S. K., & Kairon, M. S. (2019). Current status of organic cotton production in India. *Indian Farming, 69*(02), 09–14.

Necef, O. K., Seventekin, N., & Pamuk, M. A. (2013). A study on recycling the fabric scraps in the apparel manufacturing industry. *Journal of Textiles & Apparel, 23*(3), 286–289.

Pagell, M., Krause, D., & Klassen, R. (2008). Sustainable supply chain management: Theory and practice. *Journal of Supply Chain Management, 44*, 85–85.

Peter, T. (2002). The international market for organic cotton and eco-textiles. In R. Thompson (compiler), *Proceedings of the 14th IFOAM Organic World Congress* (p. 258). Canadian Organic Growers.

Roy Maulik, S. (2021), Handloom – The challenges and opportunities. In M. Á. Gardetti & S. S. Muthu (Eds.), *Handloom sustainability and culture* (Part of the sustainable textiles: Production, processing, manufacturing & chemistry book series, pp. 97–117). Springer. https://doi.org/10.1007/978-981-16-5665-1_5, Print ISBN: 978-981-16-5664-4.

Roy Maulik, S., Bhowmik, L., & Agarwal, K. (2016). Eco-friendly woven and printed apparel made from handloom cotton fabric. *Asian Dyer, 13*(6), 51–54.

Roy Maulik, S., Nasiruddin, K. M., & Begam, R. T. (2019). Creative digital printing on textiles. *Asian Dyer, 16*(2), 54–57.

Scoones, I. (2007). Sustainability. *Development in Practice, 17*(4–5), 589–596. https://doi.org/10.1080/09614520701469609

Shen, L., Govindan, K., & Shankar, M. (2015). Evaluation of barriers of corporate social responsibility using an analytical hierarchy process under a fuzzy environment – A textile case. *Sustainability, 7*, 3493–3514. https://doi.org/10.3390/su7033493

Sustainable materials guide //04 polyester. Retrieved 2023, October 26, from https://refashion.fr/ecodesign/sites/default/files/fichiers/Sustainable%20Material%20Guide%20Polyester.pdf

Telli, A., & Özdil, N. (2013). Properties of the yarns produced from r-PET fibres and their blends. *Journal of Textiles & Apparel, 23*(1), 3–10.

Telli, A., Özdil, N., & Babaarslan, O. (2012). Usage of PET bottle wastes in the textile industry and contribution to sustainability. *Journal of Textile Engineering, 19*(86), 49–55.

CSR Activities of Fast Fashion Brands: A Move Toward Sustainability

R. Surjit, K. S. Tarunvalavan, V. Sruthi, and H. Shanruthi

Abstract Currently, popular fast fashion brands are highly associated with capturing the market with their exclusive and fashionable product designs that abide by the latest trends and their unique short-duration production and distribution lead times. The various fast fashion brands are ruling the market by producing a large scale of fashion clothes and products for the current trend which attracts a large scale of people in the society. In producing fast fashion goods, they not only serve the people's needs but also end up deteriorating the environment. However, these features pose various sustainability challenges. From manufacturing a product till it gets disposed of, various factors affect the world in terms of sustainability. On looking into the corporate social responsibility (CSR) practices carried out by the brands, it is found that they try their best to neutralize the impact. Recently, several fast fashion brands have adopted many ethical measures to mitigate the societal and environmental damages caused by them. However, it is significant that these brands take immense interest in creating sustainable apparel and textile products with an apparent increase in awareness about sustainable practices among consumers. Even though maintaining CSR practices and sustainability doesn't allow the brands to completely take care of its environmental and societal impacts, it does have many benefits such as improving the quality of life of the workforce, their families as well as the local community, and the society at large. There are various CSR activities carried out by these brands. This chapter will discuss in detail various fast fashion brand's CSR activities and their efforts for the welfare of their workforce, society, environment, and the world. The chapter will deal in detail with how these brands' CSR activities alleviate the concerns of the public about them.

Keywords Fast fashion brands · CSR activities · Sustainability · Environment · Society

---

R. Surjit (✉) · K. S. Tarunvalavan · V. Sruthi · H. Shanruthi
Department of Fashion Technology, PSG College of Technology,
Coimbatore, Tamil Nadu, India
e-mail: rst.fashion@psgtech.ac.in

S. S. Muthu (ed.), *Corporate Social Responsibility in Textiles and Fashion*,
Sustainable Textiles: Production, Processing, Manufacturing & Chemistry,
https://doi.org/10.1007/978-3-031-61099-8_8

1 Introduction

A fast-moving world with the youth contributing majority population calls for the never-ending demand for fast fashion. People in today's world look for all possible ways to outshine others when it comes to being in trend. At present, numerous fast fashion brands across the world provide the people with the latest models of fashion products. The factors like subsequent growth in industrial development, advanced technologies, effective supply chain, logistics, and the industry's desire to make the most profit out of the consumers' needs led to the increased production of fast fashion apparel. However, not all these factors are desirable, and some of them cause tremendous load on the environment. For example, more industries mean more production, more energy consumption, more water usage, large-scale usage of limited resources, more laborers, etc. More production involves heavy and continuous production processes to meet the fast fashion demands of the people that add to the workload of employees and thus an unpleasant environment for workers. This again increases the requirement for more laborers for extensive production. The requirement of more energy as fuel for consumption during production in turn leads to the further destruction of available limited resources. All these affect either the society or the environment that surrounds the industry. Though these issues seem less chronic at present, they can pose life threats in the future which is not too far. And thus to cope with these problems, companies are engaged in mitigating these problems in many ways. One such way that is followed by the companies is the practicing of different corporate social responsibility (CSR) activities. As a part of this, each fast fashion brand tries to take responsibility for looking after its workers, society, and the environment around it which are more prone to suffer due to its production processes and technologies. In this chapter, we will look into different fast fashion brands like ZARA, H&M, Uniqlo, Eileen Fisher, and Primark to name a few with a detailed view of their CSR activities, their importance, and their ethical manufacturing processes.

2 Fast Fashion Brands

The term fast fashion can be described as selling the fashion trend garments at a rapid speed. In other words, the fast fashion brand products are mostly low- to medium-cost garments that make one more stylish. It is the one that moves quickly from catwalks to retail stores and falls into the hands of consumers like fast food. The First Industrial Revolution played a significant role in changing all the business sectors including the fashion industry in various means. Before the era of the Industrial Revolution, fashion was at a very slow pace where all the garments were made at home and in small fashion houses. The same garments were patched, reprocessed, and reused many times. It created a great impact in the fashion world both directly and indirectly, by the introduction and the rise of new textile machinery.

This paved the way for the production of ready-to-wear garments on a larger scale. During the Industrial Revolution, machines started to replace manpower in the work area, and mechanization took place everywhere. This can be considered one of the major reasons why fast fashion started to emerge. This laid the foundations of the constantly evolving production of fashion garments since 1800. The invention of the sewing machine bestowed a lightning reduction in the price and a drastic increase in the manufacturing of fashion clothing. After the Industrial Revolution, people got to know about the trends that were going on in and around their cities and started to wear clothes as their self-expression rather than wearing the same old clothes for decades. Fashion picked up its speed in the 1960s (Sara, 2018; Alex, 2022). These fast fashion clothes touch their peak stage and decline within a short period; only some will remain as exemptions. In today's fast-moving world, people are going crazy behind fast fashion to represent them as fashionable among others. Knowing the people's urge for fast fashion products, many brands took this opportunity to satisfy their needs and started to produce in bulk. In this way, people started to get high-end products at low cost in their nearby stores, thus increasing the profit of the fast fashion brands. Brands like ZARA, H&M, Topshop, and Primark all sprouted as small shops in Europe, and now they are everywhere in the world. In today's market, they are the top-selling multinational fast fashion brands that hold a large share of the market. These brands are fulfilling the need for fast fashion, but at the same time, they are polluting the environment in many ways. The impact that they create is alarming and is a threat to the environment. Industries should take responsible action to avoid polluting the environment from the production of the product till it reaches the consumer and even after disposing of it. All these have both positive and negative sides in it. Some of its advantages and disadvantages are discussed below.

2.1 *Advantages of Fast Fashion and Fast Fashion Brands*

The greatest advantage of the fast fashion is the affordability. The low production cost of the mass-produced garments allows the manufacturer and the retailers to sell them at a low price. It allows all consumers to keep up with the trend that keeps on changing. Then comes accessibility; in this era, considerably very few people are interested in going to shops for purchases, and everything can be sourced online now. Most of the customers are visiting and buying on the websites of the brand. Thereby it helps fill our rack full of fashion and trendy clothes at a low cost (Ninna, 2023). These brands closely watch the runways, catwalks, and celebrities and get to know the current trends. They reproduce the same trend and showcase the new collections to their customers. So they constantly have a flow of new trends throughout the year which also offers a wide range of varieties in the styles that catch and influence every consumer to buy the clothing. Fast fashion industries contribute a lot to the growth of the economy and to reduce unemployment. The constant production of clothes needs a huge workforce from sourcing of the fabric or yarn until it reaches the hands of the consumer like designers, merchandisers, tailors, retailer managers, and many more, thus creating a job for the unemployed. Such workforce requirements reduced the poverty rate in many countries especially in China during 1981–2010 (Nadine, 2022). Issues like the first copy of products and design infringement can be reduced in the case of fast fashion brands by giving opportunities for emerging designers where they can tie up with the brands and bring out new ideas and trends. The fast fashion brands lead to product development with quick turnaround times and also increase profit margins for the sellers with reduced promotional activities.

2.2 Disadvantages of Fast Fashion and Fast Fashion Brands

Fast fashion also has many disadvantages and drawbacks. The major disadvantage is the environmental impact that it creates. They are often considered to be the most polluting ones in the fashion industry after wet processing. Fast fashion pollutes the environment in all possible ways. It is now widely considered as a *disposable fashion*. All fashion items have a very short time to sustain in the fashion world once a new trend comes, and people just discard or replace that with brand-new products. This decreased lifespan of garments, rapid change in the market, and people's buying behavior lead to a great dumpyard of garments every year. Since the brands have to give trendy garments at a rapid pace and low prices, most of them do not consider the quality and sustainability of the products that they produce.

On looking at the supply chain process of the fashion goods, it stands at the third place in polluting the environment due to the emission of greenhouse gases. It is expected that the emission will increase by 50% by the year 2030 if the same scenario continues (Nikita, 2022). The fashion brands are producing twice the times of goods that they were used to produce in the early 2000s so the waste during preproduction and postproduction also increases at the same rate. Even though the waste generation is high, the recycling of the waste remains too low. Fast fashion brands are now taking effective measures to reduce the environmental impacts that they create by performing CSR activities and promoting sustainability.

3 CSR and Its Activities

Corporate social responsibility is a self-regulating activity that assists companies to be socially accountable to not only their stakeholders and the public but to themselves too. When a particular company is said to have its own CSR practices, it gains the so-called 'corporate citizenship' tag. Moreover, opting to perform CSR practices or activities will be an eye-opener for companies to gain consciousness of the intensity of impact they hold on society which includes economic, social, and environmental aspects. Apart from the benefits that CSR yields to the environment and society, it in turn promotes a positive brand image for the companies practicing them. Programs that promote corporate responsibility are also found to raise the morale in the workplace. Being accountable to itself and its stakeholders is the first step toward social responsibility for a company. However, the companies that adopt CSR activities are those that often grow their business to a point beyond which they can lend hands in giving back to the society. Thus, more often large corporations claim it to be their cup of tea. Because the more successful a corporation is, the more initiatives or responsibilities it could take up to set standards of ethical practices as a benchmark for its peers and competitors.

The different CSR activities are categorized into the following types:

- Environmental responsibility is rooted or instilled to preserve mother nature. This involves steps to reduce pollution, waste, emissions, and consumption of natural resources.
- Ethical responsibility is rooted in acting ethically and fairly. This includes fair treatment of all customers irrespective of their social background and positive treatment of all the workers and ensuring guaranteed benefits for them.
- Philanthropic responsibility challenges the companies on how they make their contribution to the society. This includes donations to charities or non-profit organizations and sponsoring fundraising events, etc.
- Responsibilities to empower women involve steps taken by the companies to empower women laborers by ensuring them with equal career-based opportunities, special training programs, and encouraging more women entrepreneurs.
- Awareness-based programs are a part of CSR activities performed by different companies. They come forward with initiatives like different projects and welfare programs that would benefit the society.
- Recycling initiatives include measures taken by different brands to track and get back the old products from their customers and recycle them to further make products out of these recycled materials.
- Partnership-based responsibilities involve initiatives taken by different companies to make collaborations or tie-ups with organizations that work for social and environmental well-being. They coordinate and achieve their CSR goals.

All these CSR activities, in general, are for the community. However, it helps the company build a stronger bond among corporations and employees. It also boosts morale and instills in employers and employees the feeling of being more connected to the world that surrounds them. Apart from positive impacts on the planet, companies that work on corporate social responsibility activities also benefit from various other factors such as remarkable brand recognition, better investor relations, efficient employee engagement, and proper risk mitigation. CSR adds to the companies' positive social value that can help them outshine their competitors in the market. On the whole, CSR initiatives will help the nation achieve its sustainable development goals and in turn transform the country for its betterment as well.

3.1 Importance of CSR

As a corporate company, each one of them should be aware of their activities that cause both positive and negative impacts on the employees, environment, and society. Nowadays consumers and even shareholders are expecting the brand to be socially responsible in the most possible ways. Corporate social responsibility is one of the wise ways to make their brand sustainable. Doing CSR activities and showing the better side of the brand will promote sustainability practices and thereby create awareness among the public. In that aspect, all fast fashion brands are trying to promote their CSR activities on various platforms, especially on social media. Doing this not only showcases their activities but also greatly increases the brand's image. CSR activities also include the welfare of the employees by catering to their needs and taking care of their physical and mental health in the workplace. The importance of CSR is discussed below:

1. Brand image: By undertaking CSR programs, the brand's image will increase as a by-product in the consumer base. When a consumer is satisfied and completely knows about the company's efforts and measures that are being taken, it will create a great impact in his or her mind. When a brand is doing some activities related to public welfare, it will create positive word of mouth, and the name of the brand will also be heard and known by many people. It would lead to increased sales (Goodera, 2023).
2. Cost efficiency: Sustainability and cost efficiency are always opposite to each other. It is difficult to sell a sustainable product at a cost-efficient price. The cost increases when a product has to be made sustainable. It requires more research to produce cost-efficient sustainable products. As a part of CSR activities, companies can promote sustainable products by selling them at lower prices so that the consumers will develop an interest in and like them. If not for CSR, such activities will not be carried out, and so CSR is important for the betterment of the community, society, and nation.
3. Risk management: It is important in the risk management sectors. By knowing and responding to the risks, a company can reduce their impacts, and further they can improvise their methodologies by performing CSR programs and practices.
4. Employee retention and engagement: Nowadays employees are not only seeking a job but also looking to work in a better company that has a good image. The volunteering activities improve employee satisfaction and encourage them to be loyal to the company (Hency, 2019).

One of the best examples of CSR being important in creating a brand's image can be explained by the partnership program by Adidas with Stella McCartney and Parley. The collaboration with McCartney was about producing very little waste, and with Parley, it was about the design of shoes from recycled plastic collected from the ocean. Both met a huge success and increased the brand image and also made them a responsible brand that is concerned about their society (Ryan, 2019).

Another example that can be quoted is a Los Angeles-based denim brand that produces a sustainable fabric that needs very little water for its production and wet processing. When this was publicized through a social media campaign, it received a great welcome in the industry, and the customers were highly satisfied with the eco-friendly clothing.

4 CSR Activities of Fast Fashion Brands

With the development of fast fashion grows the sense of social responsibility among society and within various brands. However, the responsibilities of all these brands are categorized into the following concepts toward which they work (Rima, 2022).

4.1 Human Rights

The well-established fast fashion brand, ZARA, has its stake in social responsibility as it ensures an ethical and transparent supply chain and that all the workers under this supply chain are guaranteed fair wages. The zero-tolerance policy of ZARA toward forced labor and child labor is significant as well. Moreover, ZARA is committed to ensuring healthy and safer working conditions for workers in factories. Adidas engaged itself with its second-tier suppliers to make sure that the company's practices are well aligned with the fair labor practices as given by the International Organization for Migration Rules. The very popular luxury brand, Gucci, is also said to have formatted its supply chain in such a way that it promotes respect for human rights. It has also put forth employment standards that involve trade union agreements and principles for worker's protection system and thus respect worker's rights. Apart from these, the company guarantees the right to collective bargaining and strictly prohibits all forms of discrimination, child labor, and forced labor. Patagonia's Fair Trade was their first step which was designed to ensure living wages in their supply chain. In general, the lowest-paid people are those who are employed in the apparel industries. Patagonia, however, does not own any such factories which make them least responsible for how much workers receive. However, it has managed to improve their lives through various tangible benefits through their Fair Trade. A premium amount is being paid for every product made in a Fair Trade-certified TM factory. The corporate social responsibility goals of Eileen Fisher were created back in 1997. It aims to empower and support various communities around the world both economically and socially. They put forth this by choosing the right business partners who would agree with their labor standards, regular follow-up audits, and ensuring regular training sessions to empower the workers.

4.2 Employee Development Goals

Gucci ensures that all employees are given equal and same opportunities in terms of career development and employment. The company has also been committed to strictly concentrating on health and safety standards of the employees to prevent possible occupational diseases and accidents while at the same time improving the working conditions. ZARA has been consistent in ensuring the welfare of its employees. It provides equal opportunities that significantly promote inclusion and diversity among its workforce. It also provides various development opportunities and training for its employees. Adidas is recognized worldwide for its "employer of choice" status. The company also offers internship and apprenticeship training as a part of its learning and development program. Apart from this, the company also has dual study programs. Adidas continues its investments in expanding employment opportunities. Other leadership development programs owned by the company include providing training and opportunities for all qualified candidates irrespective of their ethnicity, gender, and sexual orientation (Marissa, 2021). Fair Labor Association (FLA) is a Washington DC-based non-profit organization that was begun in 1999 to protect workers' labor rights. Being a member of FLA, Patagonia is accountable for responsible labor practices. The employers of Levi Strauss & Co. worldwide have formed Community Involvement Teams in partnership with the local charitable units to find out the needs and create volunteering and donation opportunities. Apart from this, they worked toward encouraging employees to give their time and money to support these charitable organizations.

H&M believes in an equal society of men and women with equal voice, representation, and a society free from bias. They also hear many ethical worker empowerment goals that state about the protection of female workers from discrimination and harassment. In the manufacturing sector of H&M, around 1.3 million are working to bring out the garments in which the most contributions are made by women. They aim to achieve equal opportunities and prosperity for all working women in the fast fashion sector. They target at four different regions of supply chains such as health, careers, wages, and representation, and they have also consulted with a reputed private organization to enhance and update their gender equality strategies and policies.

4.3 Supporting Women

Levi Strauss joined hands with the United Nation's Women's Empowerment Principles, which gives guidance that empowers and advances women in communities, the marketplace, and more importantly workplace, which is under the obvious control of the company. Moreover, under the HER finance initiative taken by this company, they try to uplift female workers in the low-income category. Eileen Fisher initiated a women-owned business grant program, which in turn created a new array of businesses that are socially responsible and also serve the purpose of empowering women and encouraging them in their ventures.

4.4 Sustainable Environment

A world-famous luxury brand, Calvin Klein, is working on sustainable packaging designs. This packaging is useful and safety-ensuring to the people and communities throughout its entire life cycle. Moreover, it is said to meet all the market criteria of cost and performance as well. It involves the use of recycled and renewable source materials. It is manufactured using clean production technologies and transported using renewable energy. ZARA opted for sustainable production processes to reduce its environmental impact. By 2025, ZARA desires to use 100% sustainable fabrics in its entire collection of fashion goods. Apart from these, ZARA's exclusive "Join Life" collection reduces excessive water usage through various technologies. In recent times, with the help of waterless innovations and water reduction standards and tools, Levi Strauss has worked toward reducing water usage in the apparel industry. These innovations and services are made publicly available to others in and around the industry, and its initiative to encourage other denim companies to utilize them in their production processes marks the significant socially responsible activities of this company. Levi's even worked on the Clean by Design program, collaborating with the Natural Resources Defense Council (NRDC), to reduce the usage of chemicals, energy, and water in textile mills. Eileen Fisher, being an ethical clothing manufacturer, looks forward to infusing various sustainable practices in all aspects of their production and retailing practices. Right from opting for ethically sourced raw materials to disposing of the waste after manufacturing, all they seek is to bring in positive environmental impacts and to rule out the negative ones. Gucci perceives protection of the ecosystem and reduction of environmental impact, to not only prioritize the planet's preservation but also to ensure risk reduction. This company has taken significant steps to ensure carbon neutrality. It has also taken steps to prevent and reduce its GHG emissions and the negative environmental impact it could create. Some of the environmentally sustainable goals of the fast fashion brands are given in Table 1.

4.5 Partnership-Based CSR Approaches

As a socially conscious approach, Eileen Fisher made a partnership with NativeEnergy, and together they worked on reducing their carbon footprint and using renewable energy among different communities. As of now, they have successfully brought in green energy and replaced 100% of their store energy consumption. Along with this, with the help of carbon offset projects, they are offsetting 50% of their inbound shipping emissions (Jennifer, 2020). Levi Strauss & Co.

Table 1 Environmental sustainability goals of fast fashion brands

Fast fashion brands	Year	Goals
ZARA	2023	Usage of 100% cotton and cellulosic natural fibers
ZARA	2023	100% usage of reusable plastics
ZARA	2023	100% recycling of waste
ZARA	2030	Reduction of carbon emissions by 50%
ZARA	2040	Zero net emissions
H&M	2023	Circular design
H&M	2040	Reduction of 100% in carbon dioxide liberation
H&M	2025	25% reduction in plastic materials in the packaging process
H&M	2030	Pruning of 30% of total freshwater consumption
UNIQLO	2030	Reduction in greenhouse gases across its supply chains
UNIQLO	2030	Usage of 50% raw materials as recycled materials

partnered with Business for Social Responsibility (BSR) and thus worked on the HER project that aimed to provide proper guidance about health and access to health services to the factory workers of the apparel industry. This particular initiative was proven to be successful which led them to expand and include training on financial literacy. Calvin Klein works in collaboration with FSC, an independent non-profit organization that sets standards and certifies companies that are said to protect and guard forests for future generations. They together work toward the welfare of animals and plants and also strive to provide education, fair wages, and safety to the people who live in those forests. This non-profit organization aims to build a connection between the forests and the end users. During the COVID-19 pandemic, Adidas took its stand in ensuring the well-being of its employees and suppliers. Moreover, they collaborated with Carbon and produced up to 18,000 3D-printed face shields per week for the benefit of the frontline medical workers in the United States.

ZARA has been actively taking part in various community engagement activities. Regularly this company collaborates with different non-profit organizations and charities. Also, it extends its hands of support during times of emergencies and natural disasters.

4.6 Social Welfare Projects and Programs

Eileen Fisher made investments in launching the Iowa Wind Farms project in 2011 which indeed contributed to achieving their carbon reduction goal by replacing grid-based electricity systems with renewable energy and eliminating 92,000 metric tons of greenhouse gas pollution. Apart from this, the wind farm serves as an effective source of income for numerous families in and around Iowa. It also encouraged rural area development economically. Adidas launched a Modern Slavery Outreach Program, as a part of which it took immense efforts and steps to identify the possible risks to human rights in its upstream supply chain. Moreover, the company participated in an industry task force set up to identify, prevent, mitigate, and remediate risks of forced labor at the cotton farm level. Gucci implemented various environmental projects to curb and prevent the emissions generated during its production processes. It has also worked on projects promoting carbon neutrality. Apart from these, a unique initiative taken by Gucci is its CEO Carbon Neutral Challenge, through which it invites both apparel and non-apparel-based industries to come forward and work toward reducing greenhouse gas emissions generated during their production (Gucci, 2020). Levi Strauss & Co. is working on a sustainable supply chain which was named the Worker Well-being initiative. As a part of this approach, the company joined hands with local organizations and its suppliers to bring in force programs that focused on equality, financial empowerment, family well-being, and health. Patagonia has its own Regenerative Organic Certified programs, and more than 2000 farmers are a part of this program for cotton and food. In this particular program, they work toward soil rehabilitation and animal welfare. Along with various business leaders, a group of farmers and other experts in animal welfare and soil health formed the Regenerative Organic Alliance to work on this program.

4.7 *Recycling Initiatives*

The corporate social responsibility of ZARA includes achieving its goal of minimizing the waste generated by its production processes. In an approach toward achieving this, ZARA initiated the clothing collection program which enabled customers to return their old clothes, which were then recycled. Creating ethical and long-lasting clothes that enhance the responsible life cycle of fashion is the major mission on which Eileen Fisher operates. The company started a recycling program that would reduce fabric and fiber wastage. The garments after being collected were dry-cleaned and resold. As a part of this program, customers donated their "gently used" Eileen Fisher clothing for a return of five dollars along with a certificate. By the end of 2009, the company took back over 1.6 million pieces of its clothes by reselling and repurposing its used products as per its "Renew" and "Waste-No-More" concepts. The usage of recycled source materials by Calvin Klein in producing their packages is also a significant and considerable initiative. Doing so did not only benefit the environment but also helped the company save around 40% of its production expenses.

4.8 *Other Socially Responsible Activities*

Adidas donated up to 3.25 million dollars to the World Health Organization's COVID-19 Solidarity Response Fund and almost 1.3 million dollars to the China Youth Development Foundation during the pandemic period. Gucci promotes sustainable management of the supply chain that involves precious stones and metals in its products. It follows the RJC Code of Conduct as per the Responsible Jewellery Council (RJC). The usage of minimal materials for packaging tops the list of steps any company could take to reduce its carbon footprint. Calvin Klein, by reducing the layers of packaging, brought down its shipping weight by 32% which in turn saved almost 20 million US dollars per year. The foundation of Levi Strauss strongly believes that people affected by HIV/AIDS should not only be guaranteed critical medical services but also deserve to be treated with proper respect and dignity. As a part of a community support-based initiative taken by this company and the Levi Strauss Foundation, they have donated more than 60 million dollars in grants to HIV/AIDS organizations in up to 40 countries across the world.

5 Ethical Apparel Manufacturing

Being socially responsible is all about sustainability, ethics, sincerity, and efficient handling of resources and is mainly a long-term goal. Ethical apparel manufacturing has the purpose of manufacturing garments that are relevant and apply to every environmental and social responsibility. In this section, an attempt is made to look at how fast fashion brands can do ethical apparel manufacturing and promote CSR.

Beyond the horizons of what an apparel manufacturing firm would obtain in terms of financial statements, many other key elements help in the firm's success. The morality of the apparel organization is very significant in the current textile market irrespective of their scale of production and profitability level. This not only applies to apparel industries but to all industries that are on the verge of making environmental and societal impacts. Considering ethical behavior during garment manufacturing intensifies the faithful relationship between the buyer and manufacturer and also puts heavy reliance on apparel business relationships. Ethical apparel manufacturing practices ensure business integrity, credibility, and growth of the manufacturing country's economy (Rathnayake & Karunasena, 2016).

When it comes to fast fashion, products of those sorts became popular in the market because of their cheaper prices and faster manufacturing. Retail chain stores such as ZARA and H&M are synonymous with such fast fashion ideas and are big giants in the fashion industry. Nowadays, they are facing a lot of demerits such as the usage of cheap raw materials, unsustainable manufacturing practices, and abusive labor practices. Many fashion-based critics claim that the garments made based on fast fashion are manufactured in substandard conditions and poor wages of laborers. Usually, fast fashion goods limit the number of uses after the purchase by the consumer. The brands target young people and teens who tend to use the new garment only once or twice. That is what fast fashion is all about and is the major reason why it is also called disposable fashion. Thus it heartens the "throw-away tendency," and hence the manufacturers should pertain to the existing situation of such disposals and focus on recycling them. They also strictly contend that the manufacturing is done on an unsustainable basis, and they condemn the production of fast fashion goods. So fast fashion brands need to watch out for their ethical behavior in the apparel manufacturing process. Even though most of their apparel is outsourced and produced in other countries, they should ensure ethical manufacturing practices which could affect their credibility and market growth shortly (Mary, 2021). The ethical apparel manufacturing has two main objectives. They are sustainable manufacturing and worker empowerment.

5.1 Sustainable Manufacturing

Around nine billion people are currently residing on this planet; all of them cannot follow the proper ecological standards, and they often exceed them. To increase the standard of living of our people in the world, we are confronted with the challenge of changing the existing standards of people without affecting our ecology (Seliger et al., 2008). So every manufacturing firm is in the race to use fewer resources efficiently with more productivity to create a potential medium for a sustainable society, and it is the ideal way to be ethical for ensuring the fine living of our upcoming future generations. Sustainability has been the buzzword for the past few years in the textile and fashion industries. It is imperative in the human activities, especially in the manufacturing sector. Generally, sustainability means satisfying the current needs without compromising the needs of upcoming generations. Although it has been widely accepted by everyone as a common wish, there are a few complications when it comes to the area of practical application. At first, sustainability was majorly oriented based on environmental factors, but social and economic dimensions of sustainability were later given importance (Garetti & Taisch, 2012).

Fast fashion products contribute a major percentage to the global environmental crisis. The major impacts of fast fashion products are the disposal by consumers and environmental degradation during production. According to the United Nations, the fashion industry now ranks the second most in pollution (in all bases such as water, air, and land) of all the manufacturing industries. The waste disposal by fast fashion brands contributes to a multitude of unsustainable environmental impacts, including deterioration of the quality of air, water, and land on this planet. There is a significant rise in the popularity gained by the fast fashion industry which leads to environmentally polluting supply chain operations. Water pollution includes 92 million tons of wastewater produced per annum, and consumption per year also is around 79 million tons of water (Niinimäki et al., 2020). Fast fashion is meant to have cheaper prices of garments for short fashion cycles to generate large profit margins. If the production is stimulated in such a way, then the negative impacts of those products on our environment would be enormous, and that is exactly like the present scenario of unsustainable practices (Nadine, 2022). The carbon emissions due to the manufacturing of such goods are equivalent to emissions released by an airplane traveling around 5000 km. Fast fashion produces 10% of total global carbon emissions per annum. The dark side of fast fashion comes to light when we think about the lead time of the fast fashion garments. Fast fashion brands such as ZARA,

H&M, and Forever 21 typically take less than a month to go through such a supply chain which has stringent lead times. So the "strict lead time" policy produces more waste and more usage of water. Plastic micro-fibers enter our natural water bodies, and they cause severe deterioration to the habitats of water-based living organisms (Rashmila, 2023). Fast fashion products are sold cheaply. To reduce the direct cost of manufacturing and the price of the product, they mainly use only synthetic materials and fibers which are very cheaper compared to organic raw materials. A major percentage of plastic micro-fibers are released by synthetic materials when washed. The polyester-based fibers mostly liberate many plastic micro-fibers that contaminate water bodies. The fast fashion leads to excessive production and usage of clothing. Hence there is a need for ethical manufacturing and sustainable supply chain management.

5.1.1 Sustainable Supply Chain Management of Fast Fashion Brands

In the fast fashion industry, there is a need for sustainable supply chain management (SSCM) and eco-friendly manufacturing. There is a rising concern about corporate social responsibility and the sustainability in supply chain management (SSCM) will help these brands be more socially responsible. Sustainable supply chain management is the management of the supply chains of manufacturing a product integrated with social and environmental sustainability (Koksal et al., 2017). There are many practical difficulties in the application of SSCM in the fast fashion and textile industry. The arduous problem in the fast fashion industry is to watch how manufacturing from individual supply units can be ethically procured and accounted for the garment production, transport, purchase, after-use, and disposal (Beard, 2008). Macchion et al. (2018) suggested several major strategic approaches to sustainable supply chain management in the fast fashion industry in which they could watch an ethically manufactured product in a transparent apparel supply chain. At first, the conceptualization process of the SSCM should be done properly for traceability and transparency in the fast fashion, apparel, and textile industry for building a good SSCM. Therefore, sustainable supply chain management is one of the key factors that is lacking in the current fast fashion industry. So fashion journalists and researchers suggest having an SSCM for ethical apparel manufacturing practices.

5.1.2 Worker Empowerment

As discussed earlier, the fast fashion industry relies upon the low cost of the product and increasingly relies upon the low wages of the workers. The second main objective and one of the key factors affecting ethical apparel manufacturing is the worker's empowerment. In a year, billions of people purchase fast fashion goods all around the world. According to a study by a non-profit organization, there has been an increase in 60% of fast fashion consumption in the last two decades. The apparel manufacturing industry is the most labor-dependent industry. They are mainly based on labor and workers as every piece of garment is man-made along the enormous and long supply chain. So there are millions of apparel workers involved in garment-making process. Most of the workers are females. Fast fashion products and manufacturing orders are given by the brands to lower-income countries such as Bangladesh, India, China, and Vietnam to gain better margins by having low-cost labor in the production process. The exploitation of labor is mainly due to the globalization of apparel supply chains. It is said in some cases that the standards of working practice are not maintained in the working area and every minute a female worker suffers from an unsafe workplace environment (Jacklyn, 2023). For the fast fashion garment workers, the rights of the laborers are violated sometimes. Hence, to avoid such issues, workers must be empowered more in the fast fashion industry, and CSR practices should aim at worker empowerment.

6 Consumer Perception

Consumer perception and behavior are some of the critical factors to be considered when studying the corporate social responsibilities of any fast fashion brand. Good and sustained practices of CSR enhance customer loyalty and customer attachment. The research was done to find the consumer's minds to cast out the impact on consumer behavior relating to their loyalty, preferences, and attachment to their preferred fast fashion brand (Chaudary et al., 2016). Consumers are the significant stakeholders of fast fashion brands, and it is imperative to value the consumer's perception of these brands. There is an increase in concern among consumers toward the environmental and social aspects of sustainability. Modern consumers of fast fashion brands are well aware and highly knowledged about the environmental and social concerns of manufacturing a fast fashion garment. Competitive conditions are more vital between fast fashion brands. Despite every other competitive index, corporate social responsibility and morality play a special role. So, research findings have been found indicating the positive and negative attitudes of customers, and it also focuses on the purchase behavior of the consumers (Vahdati et al., 2015). It is important to note that fast fashion brands should know how consumers perceive their CSR activities and sustainability goals. They should also analyze consumer behavior after CSR. The fast fashion brands lack a tool for measuring consumer perceptions of CSR. They should develop business models to conceptualize the method of measuring consumer perceptions of CSR. Hence, consumer perceptions play a key role in determining the effectiveness of CSR activities and ethical goals planned for future benefits and to attract customers.

7 Scope of CSR

The CSR practices not only help in the promotion of the brands, but in turn, they provide benefits in many ways. Since the emergence of the concept, it started to spread its wings wider and wider. The CSR activities of fast fashion brands nowadays are much more active than they used to be before; they are trying to fulfill the fashion needs while at the same time bringing a positive change in the society. One of the responsibilities of CSR is informing the customers about the product in all aspects from sourcing to marketing. The demands of the customers have changed from simply getting the products, to buying and using the products that are less harmful to the environment. It is undoubtable that even in the upcoming years the concern about the sustainability aspect of the product will rise. Manufacturers must take full responsibility for developing effective CSR programs and campaigns that take them toward a sustainable world. A successful CSR activity can bring out a positive change in society. Corporate social responsibility can never be just an option for producers and manufacturers. The environmental goals are now focusing on the cradle-to-cradle system in which no product completely ends up as waste. The international brands producing high-end garments have the liability to say about their CSR activities to their consumers. The well-established brands can further take care of their employees in terms of working conditions, incentives, medical facilities, etc. Ecolabels and eco-friendly certificates for the product have a greater concern. CSR practices related to labor welfare play a major role. The government also helps undertake CSR initiatives in various ways, thus providing a better environment. Hence, it indicates that there is a huge scope for CSR activities in the future too.

8 Conclusion

In summary, fast fashion garments have become very popular among our youngsters as they comprise the major population of this planet. Fast fashion industry giants ZARA, H&M, and Uniqlo have been taking many corporate social responsibilities into business moves and strategies for a sustained future, and consumers are also getting to know many CSR activities and other unsustainable environmental activities of such fast fashion brands. Corporate social responsibility generally increases customer retention and loyalty in the consumers. CSR is often correlated with the sustainability aspects, especially environmental sustainability. Moreover, corporate social responsibility enhances the business resources efficiently and the goals of an apparel firm, thereby ensuring the ethical apparel manufacturing and sincerity of the manufacturers. The fast fashion brands are presently working on several CSR activities such as protecting the human rights of the workers and employee development goals that focus on higher standards of working conditions. Recycling consumers' garments also plays a part in the CSR activities of the fast fashion brand. Brands like Eileen Fisher get used garments, while certificates and money are provided to the customers when they return such garments. There are several impacts on the environment due to cheap materials and harsh working conditions of the employees in a factory. Environmental sustainable measures and goals have been set by the concerned fast fashion brands to find some possible remedies for the deterioration that their manufacturing caused. Consumers are also a part of these unsustainable practices as they dispose of garments after one or two uses which leads to waste landfills. Sustainable supply chain management plays a major role in building a sustainable business for fast fashion brands. It is also significant that the consumer behavior of the CSR activities of the fast fashion brands is shifting progressively. It is important to also see how the purchase behavior is changing based on consumer perception. If the brands continue to develop the above-discussed CSR activities and sustainable goals, they have a very good chance to sustain in the near future, thereby leading to the well-being and prosperity of the living beings on this planet.

References

Alex, A. (2022). The truth about the industrial revolution and fashion. https://www.panaprium.com/blogs/i/industrial-revolution-fashion2022. Accessed 23 Oct 2023.

Beard, N. D. (2008). The branding of ethical fashion and the consumer: A luxury niche or mass-market reality? *Fashion Theory, 12*(4), 447–467.

Chaudary, S., Zahid, Z., Shahid, S., Khan, S. N., & Azar, S. (2016). Customer perception of CSR initiatives: Its antecedents and consequences. *Social Responsibility Journal, 12*(2), 263–279.

Garetti, M., & Taisch, M. (2012). Sustainable manufacturing: Trends and research challenges. *Production Planning & Control, 23*(2–3), 83–104.

Goodera. (2023). Importance and scope of Corporate Social Responsibility. https://www.goodera.com/blog/importance-and-scope-of-csr. Accessed 23 Oct 2023.

Gucci. (2020). Gucci corporate sustainability and responsibility policy. https://equilibrium.gucci.com/wp-content/uploads/2020/12/03-PLC.MCS_.CSR_01_Corporate-Sustainability-Resp.-Policy-_web_en-2020.pdf. Accessed 23 Oct 2023.

Hency, T. (2019). What is the importance of CSR? https://thecsrjournal.in/what-is-the-importance-of-csr. Accessed 23 Oct 2023.

Jacklyn, M. (2023). The impact of fast fashion on garment workers. https://goodonyou.eco/impact-fast-fashion-garment-workers/. Accessed 23 Oct 2023.

Jennifer, C. (2020). Eileen Fisher's corporate social responsibility success story. https://native.eco/2018/06/eileen-fisher-responsibility-success-story/#:~:text=Eileen. Fisher's Corporate Social Responsibility Goals, practice toward sustaining our environment. Accessed 23 Oct 2023.

Koksal, D., Strahle, J., Müller, M., & Freise, M. (2017). Social sustainable supply chain management in the textile and apparel industry – A literature review. *Sustainability, 9*(1), 100.

Macchion, L., Da Giau, A., Caniato, F., Caridi, M., Danese, P., Rinaldi, R., & Vinelli, A. (2018). Strategic approaches to sustainability in fashion supply chain management. *Production Planning & Control, 29*(1), 9–28.

Marissa, D. (2021). Adidas Corporate Social Responsibility (CSR) and Sustainability. https://www.thomasnet.com/articles/other/adidas-csr/. Accessed 23 Oct 2023.

Mary, M. H. C. (2021). What is ethical fashion manufacturing? A guide on the basics. https://blog.makersvalley.net/what-is-ethical-fashion-manufacturing-a-guide-on-the-basics. Accessed 23 Oct 2023.

Nadine. (2022). Disadvantages of fast fashion. https://zerowastememoirs.com/disadvantages-of-fast-fashion/. Accessed 23 Oct 2023.

Niinimäki, K., Peters, G., Dahlbo, H., Perry, P., Rissanen, T., & Gwilt, A. (2020). The environmental price of fast fashion. *Nature Reviews Earth & Environment, 1*(4), 189–200.

Nikita, S. (2022). Fast fashion pollution and climate change. https://earth.org/fast-fashion-pollution-and-climate-change/. Accessed 23 Oct 2023.

Ninna, K. (2023). The advantages of fast fashion: A closer look at its benefits. https://medium.com/@hcnxybxby/the-advantages-of-fast-fashion-a-closer-look-at-its-benefits-9fc26ea365c5. Accessed 23 Oct 2023.

Rashmila, M. (2023). Detrimental effects of fast fashion. https://earth.org/fast-fashions-detrimental-effect-on-the-environment/#:~:text=The%20environmental%20impact%20of%20fast,amounts%20of%20water%20and%20energy. Accessed 23 Oct 2023.

Rathnayake, U., & Karunasena, G. (2016). Review of strategies to improve workplace safety through ethical climates. http://dl.lib.uom.lk/bitstream/handle/123/17251/REVIEW%20OF%20STRATEGIES%20TO%20IMPROVEWORKPLACE%20SAFETY.pdf?, sequence=1. Accessed 23 Oct 2023.

Rima, K. (2022). 7 Fashion brands with strong CSR programs. https://sociallystrategic.biz/fashion-brands-with-strong-csr-practices/. Accessed 23 Oct 2023.

Ryan, M. (2019). Corporate social responsibility in the fashion industry. https://www.bu.edu/prlab/2019/03/25/corporate-social-responsibility-in-the-fashion-industry. Accessed 23 Oct 2023.

Sara, I. (2018). Fashion history lesson: The origin of fast fashion. https://fashionista.com/2016/06/what-is-fast-fashion. Accessed 23rd Oct 2023.

Seliger, G., Kim, H. J., Kernbaum, S., & Zettl, M. (2008). Approaches to sustainable manufacturing. *International Journal of Sustainable Manufacturing, 1*(1–2), 58–77.

Vahdati, H., Mousavi, N., & Tajik, Z. M. (2015). The study of consumer perception on corporate social responsibility towards consumers attitude and purchase behavior. *Asian Economic and Financial Review, 5*(5), 831–845.

The Importance of Corporate Social Responsibility Studies in Textile Companies and a Case Study: Ozanteks Tekstil Company's Corporate Social Responsibility Studies

Fatma Filiz Yildirim, Sultan Aras Elibüyük, Serap Öztürk, Ozan Avinc, Mustafa Çörekcioğlu, and Aşkıner Güngör

Abstract Today, sustainability studies are very important for companies to stay one step ahead of their competitors during intense competition and branding processes. Sustainability studies are process studies that include corporate social responsibility as well as innovative practices. In its most general lines, corporate social responsibility is to emphasize that companies make plans by taking into account all economic, environmental, social, and ethical factors, protect the welfare of the society, evaluate the social effects of activities, and help in developing human values. The notion of corporate social responsibility (CSR) is an extension of managerial ethics, and it is the responsibility of the management to make decisions in a way that will benefit the welfare and benefit of all stakeholders such as employees, customers, and society. In this chapter, information regarding corporate social responsibility studies, efforts, approaches, and implementations in textile companies are given. Moreover, at the end of this chapter, a case study about Ozanteks Tekstil company's corporate social responsibility studies and efforts is explained.

F. F. Yildirim
Department of Textile, Clothing, Footwear and Leather, Shoe Design & Production Program, Pamukkale University Acıpayam Vocational School of Higher Education, Denizli, Turkey

S. A. Elibüyük · S. Öztürk · M. Çörekcioğlu
R&D Center, Ozanteks Tekstil San. Ve Tic. A.S, Denizli, Turkey

O. Avinc (✉)
Faculty of Engineering, Textile Engineering Department, Pamukkale University, Denizli, Turkey
e-mail: oavinc@pau.edu.tr

A. Güngör
Faculty of Engineering, Industrial Engineering Department, Pamukkale University, Denizli, Turkey

S. S. Muthu (ed.), *Corporate Social Responsibility in Textiles and Fashion*, Sustainable Textiles: Production, Processing, Manufacturing & Chemistry,
https://doi.org/10.1007/978-3-031-61099-8_9

Keywords Corporate social responsibility · Social responsibility · Sustainability · Ozanteks Tekstil

1 Introduction

Studies on sustainability are now crucial for businesses to stay one step ahead of their rivals in times of fierce competition and branding campaigns. Process studies encompassing innovative practices and corporate social responsibility are known as sustainability studies. The definition of corporate social responsibility (CSR) was first put forward in the 1950s and was evaluated as the obligations of businessmen and companies and creating various policies and activities and taking decisions based on these obligations. In the past, the social responsibilities of businesses were only limited to profit, contribution to the economy, taxes they paid to the government, and employment they provided. However, with the increasing awareness among consumers, the definition of corporate social responsibility has been briefly explained as a business strategy that includes business ethics and is implemented by considering the interests of companies in all groups with which they have business relations. In its broadest sense, corporate social responsibility emphasizes that businesses should protect society's welfare, assess the social effects of their operations, and contribute to the advancement of human values when making plans that take into account all relevant economic, environmental, social, and ethical factors. Corporate social responsibility increases the brand in general as well as the image of the products they manufacture. In this way, many companies are able to establish long-term relationships with their stakeholders and increase their competitive power by improving their brand image and reputation. Many studies in the literature have shown that corporate social responsibility studies positively affect the brand.

Today, the society expects companies to sell safe products, protect the environment, treat their employees fairly, pay satisfactory wages, and be honest. The idea of CSR is a development of managerial ethics, and management has an obligation to make choices that will advance the interests of all parties involved, including society, customers, and employees. As can be understood from all these definitions, social responsibility has two dimensions: the first dimension is the firm, and the second dimension is the society. The textile sector is the second most polluting sector and creates large employment in the countries it is involved in. In this respect, it becomes even more important in this competitive environment that manufacturers and textile brands carry out social responsibility studies and projects in environmental and social terms. For example, in this chapter, the corporate social responsibility projects of Ozanteks Tekstil company and the contributions of their projects to the environment, society, and employees were examined. Ozanteks Tekstil company has been operating since 1973, has 1200 employees, and is among the largest towel and bathrobe manufacturers in Denizli, Turkey. In the company, there are dyehouse, printing, weaving, confection (end-use manufacturing), knitting, and wastewater facilities. Under the leadership of its R&D center, the company organizes monthly

Kaizen days. The company, which has certificates such as CMIA, Higg Index, GRS, GOTS, BSCI, BCI, Fairtrade, SEDEX, and OEKO-TEX STeP Made in Green, is very sensitive about social responsibility activities. They use platforms such as BVE3 to monitor the environmental impacts of chemical use in the company's dye-house processes to both control the chemical inventory and reduce the discharge of hazardous chemicals. Within the scope of ZDHC, they show their support for the environment by committing not to release hazardous chemicals into receiving environments. Ozanteks Tekstil company, which cares about nature and carries out projects to leave a livable world to future generations, donates samplings regularly every year. Focusing on the development of environmentally friendly products and services in line with its sustainability goals, Ozanteks works on its activities continuously. Within the scope of "Zero Waste" projects, they separate cardboard, nylon, glass, metal, domestic battery, and food waste generated in offices and cafeterias at their source and recycle them. Ozanteks Tekstil company, which always attaches importance to its employees, provides scholarships to the children of its employees studying at university, within the framework of instructions that are reviewed every year. Moreover, this company donated the books written in Braille alphabet to Denizli School for the Blind. This company, which organizes visits to children with leukemia and the old-age home, supports not only its employees but also other members of the society. This company, which has done a lot of work on reducing water and chemical consumption, was awarded the first prize in the water management category in 2020 at the sustainable business awards. Ozanteks Tekstil company, which made it to the finals in the sustainable business awards in 2021 and 2022, continues to work.

Overall, in this chapter, corporate social responsibility studies, efforts, approaches, and implementations in textile companies are reviewed. At the end of this chapter, a case study about Ozanteks Tekstil company's corporate social responsibility studies and efforts is reported. If to be discussed in more detail, this book chapter firstly will give information about the definition of the term "social responsibility." The benefits and drawbacks of carrying out "social responsibility" will then be covered. Afterward, some social responsibility examples and practices carried out in Turkey will be given. Then, "corporate social responsibility" concept will be introduced. Then, examples and information will be given from the projects and studies carried out by some important textile companies that have a say in textiles around the world. Finally, a case study regarding the corporate social responsibility activities and practices of Ozanteks Tekstil, one of the important textile production companies, will be presented.

2 Concept of Social Responsibility

Companies that take corporate social responsibility seriously achieve significant gains. Companies that take this issue seriously increase their brand value and therefore their market value. Today, it is known that consumers evaluate a product not only with its quality and material value but also with the social responsibility activities of the enterprises that produce this product. These changes reveal the fact that businesses today have a social mission as well as an economic one. In an age where it is imperative to make a difference to achieve competitive advantage, the social responsibility awareness of businesses is a very important differentiation criterion (Güzelcik 1999: 220). In this context, the meaning and scope of the notions of social responsibility and CSR are discussed first.

The Turkish Language Association defines "responsibility" as undertaking one's own behavior or the consequences of any event that falls within one's jurisdiction, responsibility, and accountability (www.tdk.gov.tr; 20.10.2023). Responsibility arises when an individual assumes the consequences of his/her own words and actions or behaviors that fall within his/her jurisdiction. Although the sense of responsibility is a personal feeling, it becomes meaningful only within a social structure. This feeling is formed as a result of culture, and culture is an important element that enables societies to make sense of the way they live and make sense of life. Culture, which is formed as a result of social relations, determines the duties of people living in that society toward each other and their relationship styles and gives them the responsibility to act accordingly (Metin, 2006).

Social responsibility, alternatively, is explained as the voluntary activity of businesses acting according to the values and norms of the society, without expecting any direct benefit in return. Researchers define social responsibility as the totality of the obligations of businesses to determine and implement policies, make decisions, and carry out activities in accordance with both their own objectives and social values and expectations and to improve the life of society (Özgener 2000: 136). Social responsibilities are related to an enterprise's pursuit of a business plan and policy according to economic and legal conditions, business ethics, the expectations of people and organizations within and around the enterprise, and making people happy and happy (Eren, 2002).

Using an alternative definition, social responsibility is explained as "the obligations of businessmen to make decisions in this direction and to manage the business by following the wanted paths from the point of the values and objectives of the society" (Bayrak, 2001).

Social responsibilities of businesses have become an issue that is frequently highlighted in modern management understanding. Today, the society has increasingly started to pressurize on businesses to carry out social responsibility, and as a result, the chances of success of companies that are oriented only toward profit without the object of serving the society have diminished. Because they are heavily influenced by social trends while exercising their power and authority, managers are no longer able to make decisions without taking into account the human, social, political, and legal aspects of those decisions. In addition, they must take into account environmental factors that can present both opportunities and constraints for businesses. For this reason, it is an indispensable obligation for businesses that want to survive to be sensitive to the needs of the society and to protect the environment (Ölçer, 2001).

3 Advantages and Disadvantages of Fulfilling Social Responsibility

Akyar summarized the advantages and disadvantages of fulfilling social responsibility for businesses as follows (Akyar, 2008): The advantages are:

- Studies in the social field are considered important by the public and create a sense of trust toward the business.
- It enables communication with critical target audience and science and art circles.
- It increases the social value of the business.
- It brings vitality to advertising activities and strengthens relations with the media.
- Work in the social field creates a public reputation that supports it. Companies that cultivate this kind of reputation can draw clients, staff, and investors. A more favorable environment will be more beneficial to the company's future growth and success.
- A social purpose can provide a brand with a trust or value system and can significantly improve consumer perceptions and purchasing tendency (Solmaz, 2005).
- Thanks to the partnership understanding that socially responsible companies will establish with their suppliers, honesty, and fair behavior will increase and mutual trust will be ensured. Thus, efficiency in production will be ensured through solid relationships based on long-term mutual interest.

Disadvantages of social responsibility are:

- Channeling resources to social responsibility programs violate the principles of a competitive market and may deprive shareholders of economic gain. Businesses are not organizations established to perform social activities. The major goal of businesses is to enhance profits. Social work reduces economic efficiency. Social problems are not issuing to be addressed and emphasized immediately; on the contrary, they are problems that can be solved over time within the functioning and pressures of a free market economy.
- The cost of social work in businesses is reflected in the price of the goods or services they sell, which forces the society to pay higher prices because of the work done in the social field.
- If businesses spend too much time on social issues, they may be distracted from their core mission.
- Many businesses may lack the specialized staff necessary to successfully solve social problems. In this case, businesses may have to hire new staff to deal with these issues, perhaps overstaffing and ultimately overpaying.
- Social obligations may be more expensive for firms than expected. Since the cost of social work is reflected in the price of the product, companies selling in the international market may be disadvantaged in competition with companies from other countries that do not bear social costs and thus lose their markets.

Today, the society expects companies to sell safe products, protect the environment, treat their employees fairly, pay satisfactory wages, and be honest. The notion of CSR is an extension of managerial ethics and is the responsibility of management to make decisions for the welfare and benefit of all stakeholders such as employees, customers, and society. As can be understood from all these definitions, social responsibility has two dimensions: the first dimension is the business, the firm, and the second dimension is the society.

4 Some Social Responsibility Examples in Turkey

Social responsibility activities across Turkey have increased day by day. The following are some examples of the increasing social responsibility activities:

Within the scope of the "TEB's Educated Girls" project, 1250 girls will be able to continue their education with a resource of $1.6 million created by TEB (Türk Ekonomi Bankası) group of companies, TEB employees, and BNP Paribas. Thousands of homeless children will be helped with the "Izmirim Kart" project implemented by TEB in Izmir. Bilyoner.com, Turkey's first legal games of chance platform, added a new one to its social responsibility projects by launching the "Hand in Hand for Education" project with its members in order to contribute to social development and create more livable conditions for disabled citizens (Öztürk, 2010).

The opening of the Ford Otosan Elementary School, which was realized under the leadership of the Vehbi Koç Foundation with the slogan "One Hundred Percent Support for Education" and built with the financing provided by Ford Otosan, is one of the examples of social responsibility projects. One of the successful social responsibility campaigns is the "My headlights are always on, my road is always open…" campaign that Brisa, a Sabancı Holding company, has been running in Turkey with the Bridgestone brand since 1999. This campaign was deemed worthy of the first prize in the "World's Best Social Marketing Campaign" category in 2004 by the Association of Marketing Agencies (Koç, 2013).

Cummins Inc. is a conglomerate of supplementary business units that develop, produce, supply, and maintain a wide range of power solutions (natural gas, electric, and hybrid powertrains, as well as components related to powertrains). Cummins employees in Turkey actively contribute to their communities. Various initiatives, from correcting English books for TÜRGÖK (Turkish Library for the Blind) to sorting household waste, cleaning local parks, helping elderly neighbors, and participating in disaster relief efforts, demonstrate their commitment to positive change (INT 13, 2023; INT14, 2023). They also collected baby clothes for donation and embraced sustainable living by separating organic waste for home gardening. These employees exemplify a culture of compassion, environmental awareness, and community involvement that inspires others to contribute to building stronger, more compassionate communities.

OPET has been carried out "Clean Toilet Campaign." Here, OPET initiated the "Clean 9 Toilet Campaign" to promote toilet hygiene in Turkey, beginning with its own stations, as part of the Clean Toilet Campaign, one of the longest-running social projects in Turkey, which it has been carrying out since 2000. In the "Social Responsibility Projects" category, the OPET Clean Toilet Campaign took home the top honors at the 2004 IPRA (International Public Relations Association) Golden World Awards. It also took the first place in the "Social Responsibility Projects" category of the HID (Public Relations Association) Third Golden Compass competition and the "Social Responsibility Campaigns" category of the Nokta Magazine "Doruktakiler" survey that same year (Karahan, 2006).

5 Corporate Social Responsibility

The awareness of corporate social responsibility (CSR) appeared in the 1960s and as an outcome of the acceleration of social and environmental movements in those years and the pressure exerted by non-governmental organizations on businesses, the activities, activities, and social roles of businesses began to be questioned (Deren van het Hof & Çabuk, 2011; Turancı, 2021). Accordingly, issues related to the environment and natural resources, which previously did not attract as much attention as they do today, have come to the fore in parallel with the increasing environmental problems (Karalar & Kiracı, 2011; Turancı, 2021). Elements including the degree of consumer awareness, ecological sensitivity and rapid depletion of natural resources, and social awareness have led to serious alterations in the business practices of organizations. As an outcome of increasing environmental pollution and the development of environmental awareness, understanding sustainability has grown in importance for the fashion industry (Shen, 2014; Turancı, 2021).

Social responsibility, as a multidimensional concept, can be applied in many areas. Carroll's "Corporate Social Responsibility Model," which is one of the most widely used models in the literature, consists of four elements: "economic responsibilities, moral responsibilities, legal responsibilities, social responsibility (philanthropy, volunteering projects)" (Peltekoğlu, 2018; Turancı, 2021; Iraz et al. 2017). As can be understood from Carroll's model, social responsibility basically reflects an awareness and understanding that an institution or organization should have. While corporate social responsibility campaigns are the concrete outputs of this understanding, the sense and awareness of responsibility is a principle that should be visible in all actions of the organization (Turancı, 2021). It is also monitored that the concept of social responsibility is a perception that is generally attributed to businesses and is addressed as corporate social responsibility (CSR) in the literature (Hasanov, 2010; Iraz et al. 2017). The concept of CSR is an extension of managerial ethics and is the responsibility of management to make decisions and act in a way that is for the welfare and benefit of all stakeholders such as employees, customers, shareholders, and society (Daft, 2015; Iraz et al. 2017).

6 Some Examples of Corporate Responsibility in the Textile Industry

Fashion industry is a sector based on mass production depending on the technical possibilities brought by industrialization. While the ready-to-wear sector keeps pace with the rapidly transforming world, it also transforms itself very rapidly and takes its place in front of the consumer with a new fashion trend every moment. While this situation enables the latest fashionable clothes to meet consumers at affordable prices with the understanding of "fast fashion," it also points to an order in which brands create almost weekly creations. At the same time, this understanding shortens the lifespan of products, causing them to be consumed faster, and therefore constantly feeds the need for replacement (Turancı, 2021). The concept of "fast fashion" was used "to describe the rapid change in fashion in the late 1990s and to characterize the consumption pattern that some companies began to adhere to" (Buzzo & Abreu, 2019; Turancı, 2021). This understanding, which is widespread today in connection with the understanding of modern fashion, can be explained within the framework of the characteristics of traditional and modern clothing. It can be seen that fast fashion is based on continuous and uninterrupted consumption. The logic of continuous consumption has many social, economic, social, and environmental impacts. While fast fashion causes major problems in our planet, also it has "led to a huge consumption with unsustainable demand" (Turancı, 2021). Different fashion and sustainability approaches have been proposed to eliminate this situation. Sustainable fashion can be explained as "a fashion trend which involves environmentally friendly, recyclable, high quality products." Sustainable fashion aims to "slow down the fast consumption behavior caused by the fast fashion trend and ensure ecological, social and economic sustainability" (Gürcüm & Yüksel, 2011; Turancı, 2021). With the increase in environmental awareness, it is seen that the fashion industry has adopted social responsibility and sustainability strategies on a large scale (Koca et al., 2016; Turancı, 2021).

The idea of "sustainability" has gained prominence in all fields in recent years. Sustainability is often paired with and considered together with the definition of "corporate social responsibility" (Joy et al., 2012; Turancı, 2021). In fact, it would be correct to define sustainability as practices and plans made to achieve the most beneficial results with the least use of resources (Koca et al. 2016; Turancı, 2021). There are many different motivating factors that can impact a company's CSR initiatives. It is possible to list some of them as follows: increasing awareness of ethical consumption regarding environmental problems, health concerns, human and animal rights, and pressure from international and local pressure groups. Multinational companies are under pressure to reduce costs, and global business competition is intensifying. Therefore, these companies are reviewing their sourcing strategies, labor standards, and environmental policies and face increasing demands to improve their important part in developing nations' social and economic development (Cooke and He, 2010; Atılgan et al., 2012). In this context, there are five trends that are effective in the area of CSR and are thought to shape the future, such as transparency, sustainability performance reporting, environmental awareness, cooperation, and consumer participation (Atılgan et al., 2012).

Textile and apparel businesses carry out many corporate social responsibility projects. Businesses carry out these projects for reasons such as the international legal and structural regulations, an intense competitive environment, and raising consumer awareness and responding to their demands and expectations and increasing employee satisfaction levels and profits. Companies' CSR efforts are commonly molded within the framework of environmental pressures and social standards from international buyers. The sector where this pressure is strongest is the textile sector (Atılgan et al. 2012). It was monitored that the reputation of businesses that implement CSR projects in the sector has increased. In addition, customer satisfaction, brand values, and employee loyalty increase as a result of these projects (Atılgan et al. 2012). It is known that CSR initiatives assist businesses in reducing risks to their operations, obtaining funding, luring and keeping exceptional staff, building new clientele, and gaining the respect of their communities (Cooke and He, 2010; Atılgan et al. 2012).

Four broad categories typically comprise CSR activities: market, workplace, environment, and society. Under these four headings, it is possible to show environmental issues as the oldest and currently most frequently reported issues. These environmental concerns comprise waste, recycling, limits on the use of chemicals, raw material and water consumption, energy consumption and emissions, and waste (Jones et al., 2005; Atılgan et al. 2012). Corporate social responsibility programs are a powerful tool used by the Turkish textile industry to increase its competitiveness in a variety of markets. In this context, the Turkish textile industry also collaborates with European non-governmental organizations to arrange volunteer projects (Atılgan et al. 2012). Some CSR Projects in the textile sector in Turkey and around our planet are given below..

Koton The Koton brand started its sales with a 25-square-meter store opened in Kuzguncuk, Istanbul (Turkey), in 1988 and continues to work as an international fashion retail brand today. In line with the Respect for Life manifesto and guided by the UN Sustainable Development Goals, the brand has added a new one to its sustainability efforts. Ninety thousand plastic bottles were recycled, and 28 million liters of water was saved with the collection produced. Koton, Turkey's first Better Cotton Initiative (BCI) signatory brand, has used environmentally friendly production techniques in its collection designs with the understanding of "respect for water," aiming to protect every drop of water. The collection, which uses completely eco-friendly materials and production techniques, resulted in 45 tons of less carbon emissions. In the collection, leather from apple peel, organic cotton, BCI-certified cotton were recycled, and 50% water consumption-certified Lenzing EcoVero Viscose fabrics, sewing threads, and other accessories were also selected from natural and environmentally friendly products (INT5, 2023).

Koton follows policies to support sustainable development with the awareness of its responsibilities toward its employees, society, and the world. It acts with the awareness of the power of the business world to initiate and realize transformation with its sphere of influence. It has fulfilled its responsibilities for sustainable development both through the developments that fall on the share of the fashion retail/ready-to-wear sector, which is its field of activity, and through the initiatives it has taken on social issues. It has implemented the EI labor project to include women who are excluded from the labor force into the economic system (INT5, 2023). In addition, the brand, Koton, has endeavored to fulfill its responsibility by taking various initiatives to improve the social and business lives of all women, especially its women employees, who have a great share in its successful journey in fashion retail. With the awareness that development cannot be achieved without achieving gender equality, Koton develops consistent policies in this regard. The Gender Equality Policy, signed by the Koton Board of Directors, includes many important articles underlining that as an equal opportunity employer, a fair wage and career policy will be followed and improvements for women will be prioritized (INT5, 2023).

Zorlu Textiles Group (Korteks and Zorluteks) The Zorlu Textiles Group manufactures and sells home textiles, curtains, and yarn in Turkey. Korteks, one of the Group's main textile companies, and holds ISO 14001:2004 Environmental Management System certification and also Eco-Tex Standard 100 Class 1 certification.

With its inverter project, Korteks took the second place in a 2006 competition. With the systems implemented within the scope of the project, 35% savings were achieved. Korteks is sensitive about environmental activities and encourages its suppliers to participate in environmental activities. In the supplier selection and evaluation system, Korteks gives extra points to companies that have ISO 14001 and helps spread the activities. Bins for batteries, paper, plastic, metal, and hazardous waste can be found in different departments. Harmful waste collected separately from these bins is either disposed of or sent to recycling facilities. The non-hazardous waste is passed on to recycling facilities to be reused as raw substance.

In 2005, Zorluteks began a project to restore a wastewater treatment plant and manage water resources, intensifying energy-saving efforts and reducing daily water consumption. The factory has a treatment plant with a 7200 m^3/day capacity for the wastewater it produces, and all waste chemicals are treated in this treatment plant. Improvements are made to the existing chemical recipes to minimize chemical consumption. Zorluteks carries out annual tree planting activities. Another effort to increase public awareness is the use of handbags made of recycled cloth instead of plastic handbags in the factory outlet (Atılgan et al. 2012).

Mavi Giyim Sanayi ve Ticaret AŞ The company's environmental, social, and governance initiatives are detailed by Mavi, drawing from its "For Everyone" sustainability plan. Since 2020, Mavi, the top jeans and clothing brand in Turkey, has increased the scope of its sustainability initiatives to incorporate sustainability into its broader vision of industry leadership. The company's sustainability strategy is "All Blue. Everything is better. For everyone." The value chain is where transformation for a better future begins in a people-oriented Mavi culture. Mavi carries out its obligation to guarantee decent living and working conditions for its partners in business, suppliers, and employees. Mavi's strength comes from its suppliers and staff, who support their growth and share responsibilities in the company's quest for excellence (Mavi Annual Report, 2022).

Elvin Tekstil Elvin Tekstil is a curtain manufacturer from Turkey, has its own brands, and exports 80% of its production. The company has signed some agreements; under which the company's production processes becoming environmentally friendly, and the company tries to minimize waste generation. The business also collaborates with suppliers who respect the environment. Solid waste, paper, and packaging are gathered and recycled. With the improvement works carried out, the amount of solid waste, especially paper and nylon, is reduced. The company treats its wastewater and converts it into fresh water. Companies' emission measurements are carried out regularly, and the business employs eco-friendly dyeing techniques that require fewer chemicals and dyes. Glass units have been installed in the ceiling of the dyehouse facility to reduce electricity use and utilize daylight. The company cares about its employees and celebrates their birthdays and marriages with tree donations to Tema (Atılgan et al. 2012).

Yünsa The foundations of the company's sustainability approach are based on the goal of creating value for all key stakeholders by considering social and environmental responsibilities beyond creating economic value. For a sustainable working life, the company works to provide a working environment where ethical values are kept alive, mutual trust is ensured, responsibilities are embraced, achievements are appreciated, diversity and difference are respected, opinions and expectations are considered, work and private life balance is observed, and policies and practices are carried out beyond legal requirements. The company works to measure and evaluate energy consumption, make improvements in this regard, use resources efficiently, combat global climate change, and reduce greenhouse gas production. The company also works to reduce water consumption and recycle wastewater (Yunsa Sustainability Report, 2021).

Yeşim Tekstil Yeşim Tekstil company is based in Bursa, Turkey. Yeşim Tekstil manufactures knitted products, dyed fabrics, printed fabrics, apparel, and home textiles. The company, which holds Eco-Tex Standard 100 and SA800 certificates, strives to preserve the environment throughout the entire production process by implementing energy-efficient projects, engaging in environmentally friendly activities, and recycling. The company embraced Şükriye, one of the poorest villages in Bursa, and revitalized the rapidly disappearing agriculture and animal husbandry in the village. As part of the World Environment Day, Yeşim Tekstil organized various activities to draw the attention of employees and raise awareness about the importance of the environment. Being one of the first businesses in Turkey to manufacture fabrics made of recycled cotton under the brands Yeşim Recycle and Yeşim Organic, Yeşim Tekstil patented its organic fabrics. Energy-saving projects led to reductions in CO_2 emissions and water consumption. The projects realized in recent years have resulted in 20% savings in natural gas, 10% in electricity, and 25% in water. Yeşim

Tekstil also treats and discharges 10,000 tons of water per day with its in-house treatment plant commissioned in 2000 (Atılgan et al. 2012; INT6, 2023).

Bossa The company states that it works closely with its stakeholders, including the public, non-governmental organizations, and employees, to fulfill its social and environmental obligations to society, regardless of where it operates. According to Bossa, every employee has the right to work in a safe and healthy environment that upholds human dignity. Furthermore, the organization, which responsibly manages its environmental impacts, decides what needs to be improved and developed in order to reduce or completely eliminate the environmental impact (INT9, 2023).

Nike The brand has many retail stores in the USA, all of which use energy management systems that reduce energy consumption by 36%. Nike's shoe boxes have been made from 100% recycled materials for over 20 years. And also, Nike also changed to a new box structure that uses 19% less material by weight. Nike produced 50% more shoes in 2011 than it did in 2005, but the company still sent 70% less solid waste to landfills. Furthermore, since the program's founding in 1990, Nike has gathered and processed over 25 million pairs of shoes for recycling. The brand also launched a water program in 2001 to enable material suppliers to improve wastewater quality. In addition, Nike has implemented initiatives to limit the usage of chemicals involved in dyeing and other processes performed at its material suppliers and to encourage the use of green chemistry. It is known that to date, 373 Nike shoe and clothing suppliers have made a formal pledge to participate in the Nike Green Chemistry Program (Atılgan et al. 2012; INT2, 2023; INT3, 2023).

Decathlon The sustainability objectives of the Decathlon brand include a −20% absolute reduction in CO_2 emissions by 2026, based on emissions measured in 2021, and adapting and changing to meet the demands of sustainable development. The brand has team members from more than 30 countries. The three primary pillars of the plan—people development, environmental protection, and sustainable value creation—were co-written by the teams responsible for the brand, human resources, quality, compliance, and finance. When it comes to tackling climate change, the brand recognizes that it has an impact on the environment and is actively playing its part in making change happen. But it's not simple: tackling climate change presents many challenges (INT12, 2023).

Marks & Spencer (M&S) The brand has concentrated on dyeing, polyvinyl chloride, and pesticides. As part of its corporate social responsibility efforts, the company has replaced approximately 99% of chemicals and packaging that cause environmental problems when disposed of as of 2002 (Jones et al. 2005; Atılgan et al. 2012; INT1, 2010). M&S diverted four million pieces of waste clothing from landfills between 2007 and 2010. With 120 million coat hangers recycled or reused annually, it has increased hanger recycling. The brand has also increased energy efficiency in its stores by over 10% and fuel efficiency by over 20% (Atılgan et al. 2012; INT1, 2010).

Columbia The Columbia brand surpasses ample benefit plans and professional prospects by promoting healthy lifestyles and employee well-being. It is of the opinion that emphasizing the full employee will enable them to act morally toward one another, their business associates, and their communities. The company understands the importance of clean water for a healthy planet. For this reason, it collaborates with the US nonprofit Planet Water Foundation, whose mission is to combat the world's water crisis by granting access to clean water. In communities throughout its supply chain, the brand has given access to clean water to about 46,000 people thanks to their partnership.

Through field trips, classroom activities, and educational events, Columbia and the National Park Foundation have supported the Outdoor Program for Kids since 2019. This program connects students with national parks. In its 10 years of operation, the Outdoor for Kids Program has engaged over one million students with funding from various sources. The Columbia brand collaborated with GirlTrek, a global movement of Black women who use the healing power of self-care and the ancient tradition of walking to better and transform lives, in honor of the International Women's Day in 2022. Columbia also supports the life-saving efforts of the Skin Cancer Foundation by being a member of their corporate council. The mission of the Skin Cancer Foundation is to inform the public about the link between sun damage and skin cancer as well as the value of sun protection in the fight against skin cancer (Columbia Annual Report, 2022).

Gap From the standpoint of the brand, being environmentally conscious goes far beyond simply selling eco-friendly goods. According to Atılgan et al. (2012), the brand is a global retailer with the capacity to influence significant ecological issues like combating climate change and energy conservation. The company has been a part of the Carbon Disclosure Project and the Climate Leaders Program of the US Environmental Protection Agency since 2003. Within the scope of these projects, the brand follows its performance in decreasing energy use and greenhouse gas emissions. The program determines department stores that have high energy consumption and works with their managers to achieve energy savings in these stores. In one study, for example, light bulbs were replaced, reducing the electricity consumption by about 40%. In addition, at its distribution center in California, Gap installed a solar array that consumes roughly 1.9 million kilowatts annually. A program for water quality is one of the initiatives implemented in the brand's supply chains. By the end of 2009, participating in this program had become essential to doing business with Gap (Atılgan et al. 2012). Gap continues to work to reduce its solid wastes in its shops, delivery centers, and head office. It is estimated that this program will decrease the amount of cardboard waste by 57,000 tons every year (Atılgan et al. 2012). The company started the "Recycle your blues" campaign in collaboration with Cotton Inc., allowing consumers to drop off their used jeans at any Gap-participating location. The campaign offered participants a 30% discount at Gap Jeans in exchange for their involvement (Atılgan et al. 2012).

The *Under Armour* brand is built on innovation. That's why the brand pursues opportunities to embed circularity across value chains and aims to create long-lasting, durable products. It offers premium, stylish performance apparel designed for impact, longevity, and instant recyclability, while continuously striving to improve its products and reduce dependence on finite natural resources. In particular, it is focused on reducing its footprint by reducing the impact of the power it uses and the impact of the manufacturing partners from which materials and products are sourced. The transition to renewable electricity is the cornerstone of this work, and region by region, it is building the lofty ambition of a net-zero future. The brand is stronger when the diverse backgrounds, experiences, and perspectives of the teammates who work at the brand come together to drive innovation and support Under Armour's athletes and consumers around the world (INT10, 2023).

H&M As a brand, H&M is involved in a number of waste management initiatives. Waste management laws differ from nation to nation and occasionally even from area to area within a nation. Depending on the brand, all waste, including shipping packaging and hangers, must be recycled or reused. All standard consumer bags used at H&M are made from recycled materials. Using recycled materials, avoiding adding materials like lamination or stickers to products to increase their recyclability, selecting standard packaging to reduce waste, and making sure that packaging is easily separable so that it can be recycled or reused are some of the waste management strategies used by the brand. For over 30 years, H&M has been one of the largest buyers of organic cotton worldwide. The brand has a policy that its suppliers use Forest Stewardship Council (FSC)-certified products. Although the retail sector is a much lower water user than other textile sectors, H&M sets targets for its suppliers to minimize water use. The brand applies the precautionary principle, whereby it restricts chemicals in advance, even in cases of scientific uncertainty (Atılgan et al. 2012).

Lacoste The Rene Lacoste foundation supports projects around the world that give children access to sport. When Rene Lacoste chose the crocodile for his logo in 1927, he never imagined that millions of people would wear his emblem. The brand also never thought that the existence of crocodiles as a species would ever be in jeopardy. Lacoste was the first company to support the Save Your Logo campaign, which allows organisations or private businesses to donate to the protection of the animals that represent the brand, and also the first to adopt a logo on its garments. Today, Lacoste is pledging to save crocodiles, caimans, and gharials by preserving your logo and helping to conserve biodiversity worldwide. Lacoste has taken action to make sure that these illegal activities are not a part of the Lacoste supply chain because it opposes human trafficking and slavery. These actions involve making promises to vendors that they won't act in this way and that they'll abide by local laws. Additionally, we have the right to include clauses in our supplier agreements that permit Lacoste to use independent inspection agencies to conduct unannounced inspections of their facilities during visits (INT11, 2023).

Inditex Group Since 1995, two of Inditex's primary strategy variables have been environmental management and sustainability. In 25 of its locations, Inditex has an environmental management system certified to comply with ISO 14001. Every Inditex location practices recycling and selective waste management. Investments are made in renewable energy sources and cogeneration facilities. From 2007 to 2010, the company implemented an energy and environmental integration project. The brand requests research on reducing emissions by switching from diesel engines and boilers to natural gas ones, as well as the construction of new renewable energy plants in logistics centers (Atılgan et al. 2012; INT4, 2023).

Another project that was completed between 2007 and 2010 is called Inditex Pro Kyoto. It covers environmental initiatives like the use of environmentally friendly electric vehicles in logistics centers and the reuse me program for packaging (Atılgan et al. 2012; INT4, 2023). Another project is called EcoStores. For the store furniture included in this project, FSC- or PEFC-certified wood is utilized. PEFC- or FSC-certified price tags and paper bags are utilized. Recycled materials are utilized; moreover packaging and wrapping are optimized (Atılgan et al. 2012; INT4, 2023). Inditex and the Regional Government of Galicia have come to an agreement to establish a seed bank and carry out a project aimed at enhancing the forest ecosystem. Pull&Bear intended to plant 16,500 trees in Mexico and replace printed catalogs with an online version (Atılgan et al. 2012; INT4, 2023).

Lotto The brand pays attention to social issues and, above all, to the safety of the personnel involved in production. It has also always paid attention to the fair treatment of its employees. For years, the brand has been affiliated to the world and European Sporting Goods Industry Federations, which it says promote respect for human rights among its members, as well as respect for other important issues such as health and safety. The brand also believes that protecting the environment is a commitment and responsibility that a company has toward its stakeholders. All this means better use of raw materials, less waste and fewer waste materials, and harmful emissions, which have become an essential part of Lotto's philosophy and corporate production (INT8, 2023).

7 A Case Study: Ozanteks Tekstil Company's Corporate Social Responsibility Studies

The textile sector is the second most polluting sector and creates great employment in the countries where it is located. In this respect, it is even more important for manufacturers and textile brands to carry out environmental and social responsibility activities in this competitive environment. The corporate social responsibility projects of Ozanteks Tekstil (Denizli, Turkey), which has been operating since 1973, has 1200 employees and is one of the largest towel bathrobe manufacturers in

Turkey, and the contributions of these projects to the environment, society, and employees are examined in here.

The Ozanteks Tekstil company has dyeing, printing, weaving, apparel, knitting, and wastewater facilities. Ozanteks has an R&D Center and aims to design and develop innovative, sustainable, and environmentally friendly products/processes. The company has adopted the Kaizen approach, one of the cornerstones of lean culture. Ozanteks R&D center holds Kaizen Sharing Days regularly every month. All teams involved in Kaizen projects present their work to other employees in an environment attended by senior management. With the participation of their employees, they make continuous improvements to use natural resources efficiently and reduce environmental pollution (Fig. 1).

The Ozanteks Tekstil company aims to produce as little waste as possible. Both the products and the processes used to make them are green. The Ozanteks also collaborates with suppliers who respect the environment. Waste made of paper and packaging is gathered and recycled. Specialized areas are used to collect and send solid waste for recycling. Improvement activities reduce the amount of solid waste, especially paper and nylon. The company has biological treatment facility (Fig. 2). The company treats 3500 tons of water. Regular measurements are made of dust and emissions. Ozanteks employs eco-friendly dyeing techniques that require less chemicals and dyes. Furthermore, natural gas is utilized in place of fuel.

Different processes were implemented to reduce the high-water consumption in the overflow rinsing process in the company's dyehouse. With such projects, 180,000 tons/year (close to 60%) water savings were achieved. This savings means that the water needs of approximately 1700 houses are met annually.

In solar power plants (Fig. 3), 1,430,727 trees were saved, thus supporting the protection of living life and habitat diversity by reforesting an area of 28,000,000 m^2 (eight times the area of Central Park) (Fig. 4). The solar power plant prevents the annual consumption of 2,465,223 m^3 of natural gas. Another advantage of power plants is the reduction of carbon dioxide emitted to nature. Ozanteks Tekstil company prevents 20,000 tons of CO_2 emissions with its own solar power plants. It produces home textile products using its own clean energy with solar power plants

Fig. 1 Ozanteks Tekstil Kaizen Days

Fig. 2 Ozanteks Tekstil biological wastewater treatment plant

Fig. 3 Ozanteks Tekstil factory roof solar panels

with an installed capacity of 17 MW. They have invested 4 MW of solar panels on the roofs of the factory (Figs. 3 and 4).

All nations worldwide ratified the United Sustainable Development Goals in 2015 (Fig. 5). These goals aim to address major global challenges. Poverty, hunger, health, education, gender equality, clean energy, clean water, peace, and justice are some of these issues. Ozanteks Tekstil adopts these sustainable development goals in its business and supply chain activities and has developed procedures to implement them. The company's core principles include respect for human rights, women's empowerment, and gender equality. All employees are treated equally, and there is no discrimination based on identity, language, religion, political opinion, or

Fig. 4 Ozanteks textile environmental awareness

Fig. 5 Textile sustainable development goals (INT7, 2023)

ethnic origin. A zero-tolerance policy on corruption and bribery is followed. In addition, in order to fight against child labor, the company operates in line with the principle of decent work in its business and supply chain activities. Because employee health and safety are so important, trainings are used to increase awareness of these issues.

Sustainability is of great importance for every sector today, and the textile industry is also handled with great care in this context. The textile industry is a complex sector that causes environmental and social impacts at many stages from raw materials to production processes and from distribution to consumers. Therefore, evaluating and certifying textile products in terms of sustainability ensure that negative impacts in the industry are reduced and consumers are offered more informed choices.

By adopting these values, Ozanteks Tekstil aims to create a sustainable production model by using organic and recycled fibers in its production processes. In this process, the certification of the fibers and Ozanteks' acting with this awareness represents the protection of natural resources and the goal of leaving a clean environment. The company certifies sustainable fibers with certificates such as GOTS, GRS, BCI, and Made in Africa.

In addition, the company is subjected to social audits such as BSCI and SEDEX to enhance the company's social and ethical responsibilities. These audits evaluate topics, for instance, labor rights, health and safety, ecological impacts, and business management systems. Ozanteks holds the ISO 14001:2015 Environmental Management System certification, which represents an approach that respects environmental sensitivity. Ozanteks also holds certifications such as STEP and Higg certificates for the protection of environmental awareness (Fig. 6).

Preventing the detrimental effects of the textile industry on the environment and putting the best practices into practice are among Ozanteks' top priorities. Within the framework of Zero Discharge of Hazardous Chemicals (ZDHC) practice, they take care to supply only level 3 chemicals with the lowest environmental impact. In

Fig. 6 Ozanteks Textile certificates to increase its social responsibility

order to ensure the traceability and transparency of chemical inventories, they enable stakeholders to instantly view up-to-date inventory information on the BVe3 online platform.

Acting with social responsibility and sustainability awareness, the company donated books written in Braille alphabet to Denizli School for the Blind. Organizing visits to children with leukemia and the home for the elderly, the company supports not only its employees but also other members of the society. After the R&D center was established, energy-saving initiatives were implemented, resulting in lower CO_2 emissions and water usage. Comparing 2021 and 2022 total emission amounts, a total emission reduction of 25.6% was observed. With the projects carried out in recent years, savings of 20%, 25%, and 60% were achieved in natural gas, electricity, and water. The company, which has carried out many studies on reducing water and chemical consumption, was awarded the first prize in the water management category at the sustainable business awards in 2020. The company, which was a finalist in the sustainable business awards in 2021 and 2022, continues to work.

8 Conclusion

Numerous motivating factors can initiate sustainability. Among these motivators is corporate social responsibility (CSR). A brand's or company's perception of CSR can help them rise to the top of the sustainable innovation field. As it is known, social responsibility is the acceptance of responsibilities equivalent to company performance based on the profitability of the method, the well-being of the society, and customer satisfaction. In its most general definition, social responsibility is a business strategy which incorporates business ethics and considers the interests of all groups with which the business has a relationship. It is also known that the notion of social responsibility is a concept that is generally attributed to businesses and is also considered as corporate social responsibility (CSR) in literature. It is evident that corporate social responsibility initiatives are becoming more and more significant in businesses every day. In this context, as an extension of managerial ethics, the responsibility of companies and management to make decisions and act in a way that will benefit all stakeholders such as employees, customers, shareholders, and society is becoming a necessity rather than a fashion. It is evident that numerous large corporations and brands have created policies and initiatives as part of their CSR efforts. It is expected that these initiatives will likely extend to the whole textile industry in the near and long terms and that the scope and variety of these efforts will grow.

References

Akyar, H. (2008). *İşletmelerde Sosyal Sorumluluk Üzerine Bir Alan Çalışması.* (Yayımlanmamış Yüksek Lisans Tezi) Kahramanmaraş Sütçü İmam Sosyal Bilimler.

Atilgan, T., Kanat, S., & Ille, A. (2012). Tekstil ve Hazır Giyim Sektöründe İşletmelerin Kurumsal Sosyal Sorumluluk Gelişiminin Çevreye Etkisi. *Akdeniz Sanat, 4*(8).

Bayrak, S. (2001). *İş Ahlakı ve Sosyal Sorumluluk*. İstanbul: Beta Yayınevi.

Buzzo, A., & Abreu, M. J. (2019). Fast fashion, fashion brands & sustainable consumption. In S. S. Muthu (Dü.) içinde (Ed.), *Fast fashion, fashion brands and sustainable consumption* (p. 1–17). Springer.

Columbia annual report. (2022). https://d1io3yog0oux5.cloudfront.net/_dd23717e8b1245210477340f7aff112a/columbia/db/641/5779/pdf/2022+Impact+Report.pdf. Accessed on 2023.

Cooke, F. L., & He, Q. (2010). Corporate social responsibility and HRM in China: A study of textile and apparel enterprises. *Asia Pacific Business Review, 16*(3), 355–376.

Daft, R. L. (2015). *Örgüt Kuramları ve Tasarımını Anlamak*. Nobel Yayıncılık.

Deren van het Hof, S., Çabuk, D. (2011). Kurumsal sosyal sorumluluk. M. Işık, M. Akdağ içinde, Dünden Bugüne Halkla İlişkiler (p. 49–61). : Eğitim Akademi Yayınları.

Eren, E. (2002). *Stratejik Yönetim ve İşletme Politikası*. İstanbul: Beta Yayınları.

Gurcum, B. H., & Yuksel, C. (2011). Moda sektörünü "yavaşlatan" eğilim: Eko moda ve moda'da sürdürülebilirlik. *Akdeniz Sanat, 4*(8), 48–51.

Güzelcik, E. (1999). *Küreselleşme ve İşletmelerde Değişen Kurum İmajı*. Sistem Yayıncılık, İstanbul.

Hasanov, V. (2010). *Çalışanların İşletmelerin Sosyal Sorumluluklarına İlişkin Algılamalarının Örgütsel Bağlılığa Etkisi*. (Yayınlanmamış Doktora Tezi), Gazi Üniversitesi Sosyal Bilimler Enstitüsü, Ankara.

INT1. http://corporate.marksandspencer.com/documents/publications/2010/planacommitments2010. Accessed on 2010.

INT10. https://about.underarmour.com/en/Purpose/Sustainability/2023-sustainability-and-impact-report.html. Accessed on 2023.

INT11. https://www.lacoste.com/us/SocialResponsibility.html. Accessed on 2023.

INT12. https://sustainability.decathlon.com/. Accessed on 2023.

INT13. https://www.cummins.com/sites/default/files/files/reports/2013_2014_cummins_sustainability_report_full.pdf (p. 101). Accessed on 2023.

INT14. http://www.iaosb.org.tr/Haberler/Iaosb-Duyurulari/1903-karanligi-aydinlatan-bilinmeyeni-ogreten-stk-turgok. Accessed on 2023.

INT2. http://www.nikeresponsibility.com/report/content/chapter/energy-and-climate. Accessed on 2023.

INT3. http://www.nikeresponsibility.com/report/content/chapter/. Accessed on 2023.

INT4. https://www.inditex.com/itxcomweb/en/home. Accessed on 2023.

INT5. https://www.koton.com/projelerimiz. Accessed on 2023.

INT6. https://www.yesim.com/icerik/27/kss-yonetimi/. Accessed on 2023.

INT7. https://commons.wikimedia.org/wiki/Category:Sustainable_Development_Goals_English_icons. Accessed on 2023.

INT8. https://www.lotto.it/es/corporate-social-responsibility/. Accessed on 2023.

INT9. https://www.bossa.com.tr/corporate-social-responibility. Accessed on 2023.

Iraz, R., Abul, A., & Kurnaz, G. (2017). Çalışanların Kurumsal Sosyal Sorumluluk Algısının Yenilikçilik Performansı İle İlişkisi: Konya'da Tekstil İşletmesi Çalışanları Üzerinde Bir Araştırma. *Sosyal Ekonomik Arastırmalar Dergisi, 17*, 35–49.

Jones, P., Comfot, D., & Hillier, D. et al. (2005). Corporate social responsibility and the UK's top ten retailers. *International Journal of Retail & Distribution Management, 33*(12), 882–892.

Joy, A., Sherry, J., Jr., Venkatesh, A., Wang, J., & Chan, R. (2012). Fast fashion, sustainability, and the ethical appeal of luxury brands. *Fashion Theory, 16*(3), 273–295.

Karahan, Ş. (2006). *Sosyal Sorumluluk Kampanyalarının Tüketicilerin Satın Alma Tercihlerine Etkisi*. Yayınlanmamış Yüksek Lisans Tezi, İzmir: Ege Üniversitesi Sosyal Bilimler Enstitüsü.

Karalar, R., & Kiraci, H. (2011). Çevresel sorunlara karşı bir çözüm önerisi olarak. *Dumlupınar Üniversitesi Sosyal Bilimler Dergisi, 30*, 63–76.

Koç, A. (2013). Beşeri Sermaye ve Ekonomik Büyüme İlişkisi: Yatay Kesit Analizi ile AB Ülkeleri Üzerine Bir Değerlendirme. *Maliye Dergisi, Sayı*: 165, Temmuz-Aralık, ss, 241–258.

Koca, E., Öz, C., & Yildirim Artaç, B. (2016). Hazır giyim sektöründe sürdürülebilirliğin yöneticiler açısından değerlendirilmesi. *Tekstil ve Mühendis, 23*(103), 220–230.
Mavi annual report. (2022). https://www.mavicompany.com/i/assets/documents/2023/MAVI-ANNUAL-REPORT-2022(1).pdf. Accessed on 2023.
Metin, H. (2006). *Halkla İlişkilerde Sosyal Sorumluluk.* Yayımlanmamış Doktora Tezi, Selçuk Üniversitesi Sosyal Bilimler Enstitüsü, Konya.
Ölçer, F. (2001). Günümüzde Sosyal Sorumluluğun Değişen Boyutları ve İşletmeler Üzerine Etkileri. *Standard Dergisi, 40*(473).
Özgener, Ş. (2000). *İş Ahlakı ve Sosyal Sorumluluk* (Yayınlanmamış Doktora Tezi). Konya: Selçuk Üniversitesi.
Öztürk, A. (2010). *İşletmelerde Sosyal Sorumluluk Faaliyetlerinin Çalışanların Performansına Etkileri Üzerine Bir Araştırma.* Dokuz Eylül Üniversitesi, Sosyal Bilimler Enstitüsü, Toplam Kalite Yönetimi Anabilim Dalı, Yüksek Lisans Tezi, 116syf., İzmir.
Peltekoglu, F. B. (2018). *Halkla ilişkiler nedir?* Beta.
Shen, B. (2014). Sustainable fashion supply chain: Lessons from H&M. *Sustainability, 6*(9), 6236–6249.
Solmaz, B. (2005). İşletmelerin Değişen Konumuyla Gelişen Kurumsal Sosyal Sorumluluk Bilinci Ve Türkcell'in Desteklediği "Çağdaş Türkiye'nin Çağdaş Kızları" Projesinin Genel Bir Değerlendirmesi. *Selçuk İletişim Dergisi, 4*(1).
Turanci, E. (2021). Kurumsal sosyal sorumluluk bağlamında modada geri dönüşüm ve sürdürülebilirlik: Göstergebilimsel bir analiz. *Erciyes İletişim Dergisi, 8*(1), 347–371.
Yunsa sustainability report. (2021). https://www.yunsa.com/files/document/4869-sustainability-report-2020-2021.pdf. Accessed on 2023.

Corporate Social Responsibility (CSR) in Textile Fiber Producer Companies and a Case Study: Polyteks Tekstil Fiber Production Company's Corporate Social Responsibility Studies

Ayca Aydin, Sedat Kumartasli, and Ozan Avinc

Abstract Businesses that practice corporate social responsibility (CSR) make a deliberate effort to perform their operations and applications in a way that improves society and protects the ecology. Programs for corporate social responsibility are designed to provide a structure to an organization's charitable endeavors, community service, and creation of positive social value. Corporate social responsibility encompasses working conditions, social programs, and environmental impacts, as well as many target-focused areas of activity. The textile and fashion sector, one of the largest sectors in our planet, pushes the limits of natural resource use with its low-cost structure that encourages mass production and consumerism. One of the most important actors of the textile and fashion industry is textile fiber producers. Textile fiber manufacturers make great efforts to produce many different fibers and create fibers with new properties to meet the world's fiber needs and consumers' desires. Apart from these valuable manufacturing efforts, textile fiber producers and textile and fashion companies in many different parts of the world carry out many different projects and efforts regarding corporate social responsibility. In this book chapter, first, corporate social responsibility is introduced in general terms. Afterward, corporate social responsibility in textile fiber producer companies is examined and reviewed. Finally, a case study about Polyteks Tekstil fiber production company's corporate social responsibility studies and applications is reported.

Keywords Corporate social responsibility · CSR · Textile · Fashion · Fiber production · Polyteks

---

A. Aydin · S. Kumartasli
Research and Development Centre, Polyteks Textile Company, Bursa, Türkiye

O. Avinc (✉)
Textile Engineering Department, Engineering Faculty, Pamukkale University, Denizli, Türkiye
e-mail: oavinc@pau.edu.tr

S. S. Muthu (ed.), *Corporate Social Responsibility in Textiles and Fashion*, Sustainable Textiles: Production, Processing, Manufacturing & Chemistry, https://doi.org/10.1007/978-3-031-61099-8_10

1 Introduction

The history of the textile and fashion industry dates back to ancient times, and some practices in this sector may have negative influences on the environment and ecosystem. The fashion sector, which makes an important economic contribution to countries, also constitutes a large part of the country's economies with the employment and profit it provides (Tandon & Reddy, 2013). In the twentieth century and beyond, mass production with the fast fashion concept that spread all over the world has increased consumption and led to the growth of negative environmental impacts. As a result of the profit-oriented approach of companies and the changing dynamics from past to present, the understanding of cheap and fast fashion has become more widespread and has reached a point that pushes the limits of natural resource use. In fact, in the face of increasing demands with the effect of this trend in the fashion sector, the use of natural raw materials has been insufficient, and the use of synthetics has increased. While synthetic raw materials may have negative environmental impacts, the use of natural raw materials may also have many disadvantages (Metlioğlu & Yakin, 2021).

Changing consumer attitudes and the continuous increase in demand have also increased the number of brands. The fast fashion concept that emerged as a result of these changes has caused manufacturers to act profit-oriented by increasing the number of collections (UİB, 2017). Along with these understandings, continuous production in the fashion sector shortens the product life cycle (Can & Ayvaz, 2017). Therefore, the idea of sustainability has gained prominence. With the notion of sustainable fashion, fashion and sustainability are together, unlike the concept of fast consumption (Gürcüm & Yüksel, 2012). In the twenty-first century, this approach has become more widespread, and many companies have turned to sustainable solutions in response to environmental concerns (Durgun & Sarıbay, 2022; Chan, 2001; Vermeir & Verbeke, 2008). Yet the fashion industry, one of the sectors that contributes the most to environmental pollution, is still under scrutiny (Durgun & Sarıbay, 2022; Boström & Micheletti, 2016). In addition to being promoted by companies and legislation, the concept of sustainability has also become a priority for consumers (McKinsey & Company, 2022).

The concept of sustainability, which is highly related to corporate social responsibility, came to the forefront in the 1990s following the ecological concerns that started in the 1970s. The concept of sustainability, which aims to ensure the proper use of resources, minimizes environmental damage and contributes to the economy with innovative products and also examines the effects of the environment on consumer behavior (Metlioğlu & Yakin, 2021; Bianchi & Birtwistle, 2012).

Today, while continuing their activities, manufacturing companies need to focus on factors affecting the society, people, and the environment as well as profit-oriented activities. Companies that make a difference with their social and social impacts have realized the importance of the notion of corporate social responsibility (CSR). Over time, as awareness and consciousness have grown, the idea of social responsibility has gained popularity. Businesses that recognize the value of corporate social responsibility have begun to focus their efforts in this area in order to thrive in a cutthroat market. In relation to the concept of sustainability, businesses now prioritize corporate social responsibility initiatives (Tsoutsoura, 2004).

A company's mobilization of decisions that impact and improve the welfare of its social stakeholders is known as corporate social responsibility or CSR. CSR is a collection of thorough procedures and guidelines that are incorporated into every aspect of the business's operations, including supply chain, sales, purchasing, and production (Tsoutsoura, 2004).

In order to attain the best possible outcomes, the concept of corporate social responsibility is crucial when it comes to the actions of businesses and society. In addition, the compatibility of corporate social responsibility with business strategies creates a positive brand perception in the consumer market (Lai et al., 2010). In order for companies to be successful, they need to participate to the solution of social troubles beyond economic activities (Bayraktar & Cömert, 2018).

Studies and research have shown that all kinds of corporate social responsibility (CSR) activities in the fashion and textile sector have many advantages for companies. Textile companies that take part in CSR projects and continue their activities by adopting the CSR approach are more easily accepted by the society. At the same time, it is very important for companies to show that they are responsible companies in order to open up to international markets. When the fashion and textile sector is analyzed, the CSR activities of brands such as Inditex Group, Nike, Adidas, H&M, Puma, Esprit, and M&S, which are known and followed worldwide, are also effective worldwide.

Companies are expected to improve working conditions, regularly participate in social programs, respect human and labor rights, and take actions against environmental problems in order to enhance their recognition and improve employee quality. Due to factors such as these, the full adoption of a CSR approach is among the priorities for companies.

In this chapter, corporate social responsibility (CSR) in textile fiber producer companies is examined and reviewed. Then, a case study about Polyteks Tekstil fiber production company's corporate social responsibility studies and applications is reported. For this reason, in more detail, this book chapter will first focus on what the concept of corporate social responsibility (CSR) is and why it is needed. Then, the history, advantages, and disadvantages of corporate social responsibility will be discussed. Afterward, social responsibility examples and practices carried out by textile fiber-producing companies will be examined and reviewed. Here, examples and information will be given from the studies and efforts carried out by some important textile fiber-producing companies worldwide on this subject. Finally, a case study on the corporate social responsibility efforts and practices of Polyteks Tekstil fiber production company, one of the textile fiber-producing companies, will be reported.

2 Concept of Corporate Social Responsibility (CSR)

In addition to a company's financial and legal responsibilities, corporate social responsibility (CSR) is essentially a sustainable business model that consists of a series of voluntary actions taken by the organization to enhance its social and environmental performance for the benefit of the society (Lai et al., 2010; Tokgöz, 2018). When the date of the first description of the notion of corporate social responsibility is analyzed, we come across Bowen. It was in his 1953 book *Social Responsibilities of Businessmen* that Bowen first introduced the idea of corporate social responsibility. Bowen defines corporate social responsibility (CSR) as the obligation to operate in a way that is consistent with societal norms, expectations, and values (Bassen et al., 2005).

Examining the social impact of a company's activities on the society is another term for the concept of corporate social responsibility, which aims to provide social benefits by engaging in or supporting voluntary practices and is also defined as a form of self-regulation (Carroll, 1991; Del Baldo, 2015). "An obligation undertaken to improve the welfare of society" is the definition of corporate social responsibility, which addresses ecological issues while guaranteeing the welfare of the society (Engin & Akgöz, 2013; Turancı, 2021; Kotler & Lee, 2016). At the same time, CSR, which is seen as the requirements to provide social benefits and fulfill duties, is "the active role of companies as well as governments in the development and implementation of social policies" (Hoştut & van het Hof, 2020). CSR, as defined by the European Union, is the idea that businesses actively incorporate social and ecological issues into their operations and relationships with stakeholders (Engin & Akgöz, 2013; İstanbul Menkul Kıymetler Borsası, 2011). According to the definition made by the World Business Council for Sustainable Development, CSR is the continuous commitment of companies and society to pursue activities to improve living standards while acting ethically and in support of economic development (World Business Council for Sustainable Development, 2020; Gedik, 2020; Özcan, 2021).

Numerous people and organizations define corporate social responsibility in different ways. When the common points of these definitions are analyzed, CSR is seen as a management concept and strategy that explains how a company contributes to the welfare of its stakeholders, communities, and society while carrying out its environmental and social activities. Examining the ideas of sustainability and social responsibility is also necessary to gain a deeper understanding of CSR (Birleşmiş Milletler Dünya Çevre ve Kalkınma Komisyonu, 1991; Bozlağan, 2005).

According to the Brundtland report, also referred to as *Our Common Future*, sustainability is "meeting the needs of the present without compromising the ability of future generations to meet their own needs" (Birleşmiş Milletler Dünya Çevre ve Kalkınma Komisyonu, 1991; Bozlağan, 2005). The notion of sustainability, which has various definitions, is generally understood to refer to the capacity to manage ecological systems and guarantee their continued existence (Chapin III et al., 1996; Yavuz, 2010).

Many studies and reports have examined the concept of sustainability from different perspectives. In addition to environmental, social, and economic approaches, a multidimensional corporate sustainability model has been created in many different studies, and financial and corporate governance titles have been added to these titles (Aras et al., 2018).

Today, companies are scrutinized in many areas ranging from their economic activities to their sustainability activities, corporate social responsibility activities, principles, and performances. Companies that are held responsible for meeting the expectations of transparency and accountability by their stakeholders also benefit from sustainability reports (Engin & Akgöz, 2013).

When the notion of corporate social responsibility is researched, the concepts of business ethics, corporate social performance, corporate philanthropy, corporate citizenship, corporate responsibility, sustainability, and corporate governance are frequently encountered. The definitions of some of these concepts are as follows:

"The study of business situations, activities, and decisions that address issues of right and wrong" is the definition of business ethics (Goodpaster, 1991).

A framework known as *corporate social performance* deals with an organization's obligations to its stakeholders, workers, and the community. Other names for the idea of corporate social performance include social responsibility, corporate citizenship, and responsible business. It also includes standards and values that are adapted to the company's organizational models (Özcan, 2021).

Within the context of corporate social responsibility, *corporate philanthropy* refers to initiatives that address pressing issues and benefit the general public. Corporate social responsibility is the study of how an organization affects society and the environment (Özcan, 2021).

Corporate citizenship is a concept that advocates that a company has responsibilities toward other companies and stakeholders as well as its responsibilities toward society while carrying out its activities. At the same time, it brings along requirements such as complying with laws and human rights and working in cooperation with all stakeholders of the company and its employees. In addition, companies with a corporate citizenship approach gain the loyalty of employees and customers (Crane et al., 2008).

Corporate responsibility is a concept associated with the stakeholder theory. The concept of corporate responsibility, which encourages the company to be stakeholder-oriented and to communicate with all companies and parties that have an impact, was born out of the need for organizations to contribute to the welfare of the society (Özcan, 2021).

The idea of corporate social responsibility, which contends that businesses have obligations to the society and the environment, has been defined using a variety of models (Bayraktar & Cömert, 2018; Özcan, 2021). The following is a list of the corporate social responsibility models that have been extensively utilized in literature over the years:

- Ackerman model
- Carroll's model of social responsibility

- Frederick's model of social responsibility
- Davis' model of social responsibility
- Lantos' social responsibility model

According to the 2023 Sustainable Development Report, today's outputs need to create capital assets for the future, and it is important to ensure this. In the report, it is stated that sustainable capital assets are resources that will enable achieving economic, social, and environmental goals. These assets are categorized under eight main headings: human capital, infrastructure capital, natural capital, innovation capital, business capital, urban capital, and cultural capital. At the same time, the long-term destructive effects of a deterioration or deficiency in one of these assets, which work together in a mutually supportive manner, are also mentioned (Sachs et al., 2023).

The United Nations Global Compact was developed and published to encourage businesses to act in a more socially and sustainably responsible manner, with the goal of fostering a common and sustainable understanding of global development. The Compact is considered the world's most widespread voluntary responsibility project and is implemented by the United Nations Development Programme.

The ten principles disseminated by the UN Global Compact are grounded in widely acknowledged UN proclamations. The United Nations Rio Declaration on Environment and Development, the International Labor Organization (ILO) Declaration on Fundamental Principles and Rights at Work, the Universal Declaration of Human Rights, and the United Nations Convention against Corruption are these declarations (https://www.globalcompactturkiye.org/10-ilke/).

It is known that companies support their management and production processes with certificates and standards in order to demonstrate their ethical behavior and social responsibility. In this regard, companies benefit from internationally recognized certifications and announce to their stakeholders that they are making efforts to improve their internal processes. ISO 14000, SA 8000, and ISO 26000 standards can be considered as international standards recognized in the field of CSR (corporate social responsibility) (Montabon et al., 2000).

ISO 14000: ISO 14000 is an international standard introduced in October 1996 and consists of a series of standards. With this standard, it is possible for companies to have a certified environmental management system. This standard, which covers all environmental processes of the company, is used all over the world. It also includes communication with the parties involved in these processes. This standard, which is a guideline for businesses, is the most widely used standard in environmental management, positively affects the performance of companies, and is also seen as a selection criterion in supplier preference (Montabon et al., 2000).

SA 8000, or Social Accountability 8000, is a voluntary certification that was developed by Social Accountability International (SA) and is based on the ILO Conventions and the Universal Declaration of Human Rights. The standard, also known as the social accountability standard, also supports the notion of corporate social responsibility. The working conditions provided by companies to their employees not only provide benefits for worker safety and welfare but also bring many benefits for employers and stakeholders (https://sa-intl.org/programs/sa8000/).

ISO 26000 is one of the newest initiatives of ISO and is a 116-page standard. This standard states that every company and organization can be more socially responsible and that it is a standard that can be used to adopt this understanding (Castka & Balzarova, 2008). The standard provides guidance on what social responsibility is and its practices, related policies, stakeholder analysis, performance, and social responsibility principles. The communication protocol and documents linking the OECD Guidelines for Multinational Enterprises and the United Nations 2030 Agenda (Sustainable Development Goals) are documents that support the applicability of the ISO 26000 standard (https://www.iso.org/obp/ui/en/#iso:std:iso:26000:ed-1:v1:en).

3 History, Advantages, and Disadvantages of Corporate Social Responsibility

Since the 1950s, corporate social responsibility has been a topic of discussion. In the mid-twentieth century, businesses began to prioritize social responsibility in addition to providing economic benefits while growing. Businesses that have embraced corporate social responsibility have become more valuable in the eyes of the public (Macit, 2020). Since the 1960s, environmentalist approaches, women's and consumer rights, and civil society organizations have become more important to the society. The importance of civil society organizations, which became stronger in the 1990s, has gained a different dimension with the development of technology and easier access to information. In this process, it has become clear that it is not enough for companies to make money but that they need to show sensitive activities toward the society for a sustainable world (Turpcu, 2021). Today, with globalization and the continuous enhancement in the number of private companies, attention is paid to the quality of these activities rather than the existence of social responsibility activities. As a result, companies have started to allocate significant budgets for social responsibility projects and aim to ensure the continuity of these kinds of projects (Turpcu, 2021).

In actuality, the idea of corporate social responsibility (CSR) is a crucial subheading of corporate sustainability and plays a major role in maintaining corporate accountability and open governance (Engin & Akgöz, 2013). Today, businesses not only survive through economic activities but are also seen as structures with social, environmental, and ethical responsibilities toward their stakeholders and society (Carroll, 1991; McWilliams & Siegel, 2001). Examining corporate social responsibility initiatives reveals that non-governmental organizations (NGOs) play a noteworthy part in the adoption and execution of social responsibility initiatives undertaken by businesses, and they also have a substantial influence on the development and assessment of CSR awareness (den Hond et al., 2007; Turker, 2009; Burchell & Cook, 2011).

Many companies around the world, from the global scale to the small scale, are including more CSR projects in their annual reports and media platforms to increase their brand value (Lindgreen & Swaen, 2010; Sweeney & Coughlan, 2008; Maignan & Ralston, 2002). The notion of corporate social responsibility is associated with numerous benefits. In some studies, these advantages have been examined in various classes as benefits to the business; customers; competitors; employees; suppliers; shareholders; government, i.e., internal and external stakeholders; society; and the environment (Aktan & Börü, 2007). The advantages that corporate socially responsible companies can gain can be listed as follows:

- Continuation of the company's activities within the scope of corporate social responsibility increases brand and company awareness.
- With the social environment formed, the company is adopted by the society and becomes an important part of the society.
- Budget and effort spent on social responsibility activities increase profit margins.
- Strengthening the company's relationships with stakeholders, society, and governments.
- Corporate socially responsible companies becoming environmentally friendly.
- Discovering new markets and increasing customer loyalty.
- In the long term, the company will gain value for its employees and stakeholders, increasing their trust and sense of belonging to the company, as well as an increase in the number of qualified employees.
- Increased tendency of employees to teamwork through their involvement in corporate social responsibility projects and improved business performance through strengthened relationships within the company.
- Increased work performance, motivation, and morale of employees who take part in various projects carried out inside and outside the company, as they feel more valuable and their awareness increases.
- Contributing to companies with high corporate social awareness to continue their activities by respecting all kinds of human rights and to direct their priorities in this direction.
- Increasing the productivity of in-house employees (Seyitoğullari & Bilen, 2020; Krasodomska, 2013; Nalbant, 2005; Doğan, 2007).

Although corporate social responsibility has many positive aspects, it also has a number of disadvantages for companies. The disadvantages and risks can be listed as follows:

- May cause more bureaucracy and paperwork in the company.
- When the budget spent on corporate social responsibility activities is reflected in product costs, it may lead to price increases and loss of customers.
- Additional human resources may be required for the execution of corporate social responsibility projects.

- The voluntary nature of the activities and the risk that the persons in charge may not be able to perform the same performance continuously.
- The fact that reporting on corporate social responsibility is constantly changing and may be difficult to follow up (Özcan, 2021; Nalbant, 2005; Doğan, 2007; Jones et al., 2009; D'Orio & Lombardo, 2007).

When the advantages and disadvantages of corporate social responsibility practices are analyzed, it is evident that there are many positive effects and results. Particularly, successful initiatives not only move companies forward in many ways but also increase their competitiveness (Özcan, 2021; Seyitoğullari & Bilen, 2020; Atılgan et al., 2012).

4 Corporate Social Responsibility Practices in Textile and Textile Fiber Producer Companies

Although the textile and fashion sector is an intensely competitive environment, it takes part in corporate social responsibility projects and carries out these projects within the company for many reasons such as meeting customer demands, acting in compliance with legal regulations while meeting sector expectations, strengthening its relations with its stakeholders, and many other reasons (Atılgan et al., 2012; Geethamani, 2017; http://www.unglobalcompact.org/docs/networks_around_world_doc/Network_material/CSR_Report_in_English.pdf). At the same time, it is known that the benefits provided by CSR projects are very efficient for brands. When the brands that continue their activities both in our country and around the world are examined, it is seen that they carry out various projects within the scope of corporate social responsibility.

In this part, first, the CSR (corporate social responsibility) projects of some big companies selling textile and fashion products around the world are briefly mentioned. Then, the works, projects, and efforts of some textile fiber manufacturers in the field of corporate social responsibility, which is one of the main topics of this book chapter, are reviewed.

For instance, Inditex Group, which stands out both in Turkey and in the world in terms of brand awareness and accessibility, acts with serious expectations from its suppliers and stakeholders as well as the social responsibility projects it carries out. In this context, the company strives to increase the environmental awareness of every individual, from its employees to other stakeholders and society, and has been a signatory of the UN Global Compact since 2001. Inditex Group has publicized many targets such as protecting biodiversity in five million hectares of land, reducing its carbon footprint by a minimum of 90% by 2040, and reducing water consumption in the supply chain. In addition to these, it carries out many projects such as various start-up investments, the use of recycled raw materials, a second-hand sales platform to ensure the reuse of clothes within the scope of the circular economy model, and the creation of garment boxes in its stores. In addition, the company collaborates with more than 400 non-profit organizations and aims to help ten million people by 2025 while taking responsibility toward society. At the same time, the company regularly publishes sustainability reports and is recognized by the society as a corporate socially responsible company (https://www.inditex.com/itxcomweb/en/sustainability). Another example here is the H&M company. One of the most well-known ready-to-wear brands in the world, H&M draws interest from suppliers and customers with its various projects. The company, which has been regularly publishing sustainability reports for many years, is one of the brands that have been influential in popularizing the concept of sustainable fashion. It attaches great importance to waste management and recycling. In this context, it encourages the use of recycled products from consumer bags to product hangers (Atılgan et al., 2012). Despite working with more than 700 independent suppliers, the company ensures the traceability of its products and carries out the "Let's be transparent" project. With the "Let's be fair" project, which was initiated in many countries to protect the rights of its employees, the company strives to contribute to the welfare of the society under the headings such as fair wages, prohibition of forced labor and child labor, and animal welfare. One of its biggest projects is the "Let's close the loop" project, which has been running in its stores all over the world since 2013, to recycle used clothes. In this context, wearable garments are sold second-hand, reusable garments are transformed into different products such as cleaning cloths, and other garments are shredded in recycling facilities and used as insulation material. In addition, discount vouchers are given to people who bring in their clothes to encourage the community. In addition, in 2019, they launched a clothing rental service from their "Conscious Exclusive" collection at a store in Stockholm, Sweden. With the "Let's clean up" project, the company wants to prevent harmful chemicals, fossil-based energy sources, and disposable packaging and announced that they want to establish a water standard in cooperation with organizations such as WWF and Solidaridad. They also aim to halve their emissions by 2030. H&M shows that they are working on solving plastic pollution with the "Bottle2fashion" project, which collects PET bottles from Indonesian islands and turns them into polyester fibers. The brand, which has created a new and sustainable children's collection with yarns produced within the scope of this project, aims to reduce the amount of plastic pollution in the seas by 70% by 2025 (https://www2.hm.com/tr_tr/sustainability-at-hm.html).

One of the important Turkish textile companies working on this issue is Mavi. Operating in Turkey and exporting worldwide, Mavi carries out various corporate social responsibility projects. Mavi has consistently participated in the United Brands Association's campaign since 2012, providing clothing support to children in need, collaborating with non-governmental organizations to implement social responsibility projects that add value to young people, and supporting female students through a scholarship fund. The company is listed in the BIST Sustainability 25 Index and is on the "A" List of the CDP (Carbon Disclosure Project) platform. Mavi, which has received the "Most Sustainable Collection" award four times at the Rivet Awards for its collections, acts within the scope of social compliance with its suppliers and places importance on ethical codes of conduct. Mavi has also published "Diversity, Human Rights, Diversity and Inclusion, Biodiversity, Environment and Energy, Forest and Paper Products and Animal Welfare policies." It supports the Ecological Research Association (EKAD) and has been carrying out the Indigo Turtles project for endangered sea turtles since 2014 (https://www.mavi-company.com/tr/degerler/sosyal-sorumluluk). In addition, while supporting sustainability and corporate social responsibility, it has created a collection with jeans produced using recycled raw materials. It also uses organic cotton and has various memberships and internationally recognized certificates. It is a signatory to the United Nations Global Compact (UN Global Compact-UNGC) and Women's Empowerment Principles (WEPs) and carries out practices such as life cycle assessment, carbon footprint reporting, sustainability reporting, and energy management (https://www.mavi.com/sustainability/index.html). Another global textile company that has carried out important efforts in this regard is Nike. Nike, a US-based brand, is known all over the world as a sportswear and sneaker brand. The brand, which carries out various activities on its zero-carbon and zero-waste journey, shows its sensitivity to the environment with the collections it creates. Nike, which creates collections using recycled raw materials from pet bottles, contributes to reducing waste by recycling or donating used sports products and sells second-hand products of underused products. The BILL (Bot Initiated Longevity Lab), a robot-assisted system introduced in London, enables the cleaning and repair of products through personalization. Nike's newest initiative, BILL is an example of good work in the circular economy (https://www.nike.com/tr/surdurulebilirlik/hizmetler). It also works on reducing its carbon footprint, saving water, and eliminating hazardous chemicals and waste (https://about.nike.com/en/impact/focus-areas/protecting-the-planet). It also builds equitable and inclusive teams and creates a transparent and responsible supply chain. It contributes to social welfare by providing opportunities to access resources. In its impact report, the company announced that 51% of the global corporate workforce is represented by women, a fair wage policy is implemented in its companies, they continue to work to reduce carbon emissions in production, and they have been using wind and solar energy for years. It was stated that they continue to work to ensure a good future for all athletes, especially women and girls (https://about.nike.com/en/newsroom/reports/fy22-nike-inc-impact-report).

Now, in the next part of the book chapter, the studies and projects of textile fiber manufacturers in the field of CSR (corporate social responsibility) will be mentioned. Let's start this topic by talking about the work and developments carried out by Dyneema and Dyneema fiber manufacturers in this regard. The major environmental effects of the worldwide production, shipping, and use of textile products are a concern, according to Dyneema. For this reason, they are developing end-of-use products made with Dyneema®, which guarantees the availability of natural resources in the future, in collaboration with Avient company, the industry leader in ultra-high-molecular-weight polyethylene (UHMwPE) fiber. They conduct reassessments and recycling in this regard. Avient company introduced the first bio-based HMPE fiber in history, adopting a more ecologically friendly strategy. Reducing reliance on fossil fuel-based products, bio-based Dyneema® fiber ensures that renewable resources are used without sacrificing performance, protection, comfort, or durability. It is derived from renewable, bio-based raw materials. Based on life cycle assessment calculations (an internationally recognized procedure for comparing different solutions for the same functional unit in order to evaluate the ecological influence of a product throughout its life cycle), cut-resistant gloves made with bio-based materials (cut level 4 performance) Dyneema® fiber are only available per 10,000 pairs. It has 2800 kg CO_2 emissions. It is, respectively, six times lower than aramid fiber and 6.5 times lower than regular HMPE fiber (https://www.dyneema.com/about-us/latest-news/blog/dyneema-provides-innovative-and-sustainable-solutions-for-fabric-and-protective-knitwear).

SASA: Operating in Turkey and establishing itself in the international market, SASA is a raw material producer of polyester, fiber, filament yarn, polyester-based polymers, specialty polymers, and intermediate products. It acts in line with the United Nations 2030 Sustainable Development Goals and cooperates with many NGOs. In addition to the cooperation with non-governmental organizations such as the Red Crescent, AFAD, and the Child Protection Agency, the company provides scholarship support for the children of its employees studying at universities and organizes events for students in line with the requests from universities. The company aims to contribute to the industry with the Incubation Center Library established in Adana Organized Industrial Zone and provides employment within the scope of the İskur trainee training program. At the same time, SASA Mehmet Erdemoğlu Nursery School was started to be built in 2020 in order to enable employees to leave their children and to support pre-school education (https://www.sasa.com.tr/surdurulebilirlik/sosyal-surdurulebilirlik/sosyal-sorumluluk). The company, which regularly shares its sustainability reports with the public along with its social responsibility projects, continues to work on emission, energy, waste, and water management and holds various certificates. In 2023, it ranked 6th among 561 chemical companies in the Sustainalytics ESG Risk Rating. The company continues to work on corporate social responsibility and social and environmental sustainability (https://www.sasa.com.tr/surdurulebilirlik).

Indorama: Indorama Venetures is a globally recognized chemical company with a high market share in polyester production. The company publishes regular sustainability reports every year and holds ISO 9001:2015, ISO 14001:2015, ISO 45001:2018, ISO 50001:2018, and ISO 22000:2018 certificates. In order to raise awareness on recycling within the scope of corporate social responsibility, the company conducted trainings on waste sorting, upcycling motivation, and raising awareness on the 3Rs through collaborations. In order to improve community welfare, the company has more than 300 initiatives in the field of public health and more than 20 initiatives in the field of environment (https://www.indoramaventures.com/en/sustainability/social/csr#recycling-education). Bringing its stakeholders together with the events it organizes, the company demonstrates its corporate social responsibility through trainings on circular economy and recycling, support for library institutions, events organized for children, and support for those in need (https://www.indoramaventures.com/en/sustainability/social/csr).

Lenzing: Lenzing Group, a prominent fiber producer, is among the industry leaders in terms of sustainability and corporate social responsibility. According to the CSR rating, it has attained Platinum Status. The four primary CSR practices—the environment, human rights, fair labor conditions, and sustainable and ethical procurement—are all thoroughly covered by the assessment. In addition to producing fiber, Lenzing is always thinking about forest protection, humanitarian aid campaigns, and the well-being of its employees (https://www.lenzing.com/newsroom/press-releases/press-release/lenzing-awarded-platinum-status-for-sustainability-by-ecovadis).

Kuraray: Kuraray company raises awareness in the sector by introducing its products that contribute to sustainability at international fairs. Among these products, it produces polyester raw materials and fibers from used plastic bottles and also recycles plastics collected from the oceans. Along with these products, developing biodegradable materials in the sea can also be shown as one of these goals. In addition, the company has created a fund for donation activities to be performed in different parts of our planet. In this context, a support donation was made to the Turkey-Syria earthquake victims (https://www.kuraray.com/csr).

DowAksa: DowAksa, Turkey's first carbon fiber producer, has been carrying out its activities in a sustainable manner since its establishment. The company is concerned with global trends and growing customer sustainability requirements, having established a clear link between social and economic development and environmental impact. Within the scope of its Sustainability Strategy and Vision, it uses renewable energy sources to reduce its global carbon footprint. Thanks to the effective solutions they offer in the wind energy sector, they have also introduced emission reduction, waste management, and recycling processes. It shows that the products they produce will reduce CO_2 emissions throughout their entire life cycle and play an effective role in making the world carbon neutral by 2050 (https://www.dowaksa.com/sustainability.html).

Invista: By establishing a corporate charitable investment fund, INVISTA, together with Koch Industries, is advancing the development of people in need and those in the society. In addition, it has announced that the development of Nylon 6.6 GE will be in a more environmentally friendly manner at the Asian Innovation Center in Shanghai Chemical Industrial Park. The company develops and licenses different technologies to ensure emission reductions in the automotive materials supply chain (https://www.invista.com/stewardship/social).

NatureWorks LLC: NatureWorks LLC company, as a biodegradable polymer manufacturer, uses and develops the latest technologies to convert renewable carbon sources in the atmosphere into the Ingeo™ biopolymer portfolio. Working with the Ellen MacArthur Foundation, the company is working on different technologies to move away from fossil fuel raw materials and non-renewable carbon for a more circular, reusable, and sustainable society. Ingeo's production is carried out with products that are not agricultural raw materials. For this reason, the company uses sustainably grown bio-based carbon resources that are abundant locally in the regions where Ingeo™ is produced. For example, bagasse, sawdust, branch grass, or straw sugars can be used for this purpose. NatureWorks has shown that PLA collected in post-consumer and post-industrial environments can be recycled into the new Ingeo™ PLA resin using mechanical and chemical (hydrolysis) methods. For the long-term future of sustainable biopolymer production from renewable carbon, the company plans to bypass the agricultural step and directly convert greenhouse gases like methane and carbon dioxide into lactic acid (https://www.natureworksllc.com/sustainability/sustainable-feedstocks).

Kolon Industries: Kolon company ensures that the products that were not used for the fashion industry are included in sustainable fashion in 2012, when the concept of sustainable fashion was unheard of. Kolon not only produces recyclable products but also works to minimize the environmental damage caused by textile companies. Customers and other stakeholders are now demanding that fashion companies take people, animals, and nature into account and take responsibility. In addition, it is managed together with the company, employees, and stakeholders. It has established and started to implement ethical standards for basic ethical management and ethical management culture ("KOLON INDUSTRIES Code of Ethics"). Kolon company also developed the PBAT product, an environmentally friendly and biodegradable polymer, in 2021 (https://www.kolonindustries.com/en/sustainability/esg/report).

Antex: Antex company actively carries out studies on environmental protection, energy efficiency and water saving, and elimination of non-biodegradable waste. By producing recycled products, it provides a 50% reduction in carbon dioxide emissions, 40% water savings, and 45% energy savings (https://antex.net/sustainable-innovation/).

Huvis: Huvis company environmental protection activities are largely divided into three areas: developing environmentally friendly products that result in little or no harm to the ecology, saving energy, minimizing carbon dioxide in the process, and protecting the natural ecosystem with the voluntary participation of employees. Apart from these issues, it has adopted environmentally friendly approaches such as recycling wastewater, recycling waste heat, and using methane gas as fuel (https://www.huvis.com/mobile/eng/company/GreenBusiness.asp).

Owens Corning: The company works on waste reduction and recycling to create a circular economy. They continue to invest in this regard. In this context, it recycled two million tons of glass fiber. They aim to reach 35% representation of women (globally) and 22% representation of people of color (USA) in leadership roles at the company by 2030. It also continues its work on housing, improving the lives of veterans and their families, supporting educational opportunities for those in need, and promoting health and wellness (https://www.owenscorning.com/en-us/corporate/sustainability/docs/2023/2022-Owens-Corning-Sustainability-Report.pdf).

The Mitsubishi Rayon Group: It carries out water purification and industrial water and wastewater purification processes using MF membranes (microfiltration membranes) produced by the company. In recent years, at a time when there are concerns about water scarcity around the world, a successful path has been followed in wastewater treatment and desalination of sea water by using the membranes developed by the company. The "STERAPORE" hollow fiber membrane filter is used not only in industrial water filtration and many other types of water purification but also in the pre-treatment process for desalination of seawater. In this way, it provides both environmental and social benefits. Additionally, this system is included in the "CLEANSUI" series of home water purification devices (https://www.m-chemical.co.jp/en/csr/pdf/csr_report_mrc_pdf).

Toyobo: Toyobo company actively works on social contribution and community activities. It conducts business in a fair and impartial manner, without violating social norms in corporate activities, and prohibits corruption and bribery to all stakeholders. In order to maintain and strengthen environmental activities, it has introduced certification systems for ongoing environmental protection activities (e.g., obtaining ISO 14001). It prohibited in humane acts such as child labor, forced labor, ill treatment, and human trafficking and prevented all forms of discrimination, including those based on gender, nationality, race, creed, age, disability, etc. (https://www.toyobo-global.com/pdf/sustainability/report/2022/sustainability2022_e.pdf).

Eastman: Eastman company produces Naia™, a diacetate fiber that is also available as filament and staple fiber produced sustainably from wood pulp by an almost closed-loop chemical process, from 60% renewable wood pulp containing 40% acetic acid obtained from recycled waste materials, creating a more environmentally and sustainable solution. It aims to protect both society and the environment through the process (https://naia.eastman.com/).

Nilit: Nilit company reduces carbon footprint, greenhouse gas emissions, water, and other natural resource consumption in its studies. It has also accelerated the creation of new bio-based, biodegradable synthetic fibers and recyclable products. Apart from this, it has started to offer a more traceable process to employees, customers, suppliers, and partners with its work and new business practices (https://www.nilit.com/nilit-enriches-corporate-social-responsibility-programs-and-publishes-formal-sustainability-business-strategy-based-on-life-cycle-assessment/).

Radici: Radici company has introduced the Biofeel® brand, which is a biodegradable polymer produced from natural resources. The yarns produced from this polymer, produced using the oil of a bean grown in India, are used in several industries from fashion to sports and from automotive to domestic textiles and are produced from 100% sustainable resources. Radici Group uses this polymer to make their Biofeel® Eleven yarn. What's left over after the first pressing is a very potent biofertilizer that gets back into the soil (https://www.radicigroup.com/en/news-media/news/100-naturally-sourced-yarn-radicigroup-launches-biofeel-eleven-72512).

Asahi Kasei: Asahi Kasei has initiated an environmental investment process toward carbon neutrality by investing $100 million worldwide, aiming to solve problems in the energy and environmental fields. It aims to produce carbon dioxide storage, carbon management, and bio-based chemicals in the period until 2027. In this way, the company will meet the raw material and energy needs of the yarns it produces from sustainable sources (https://asahi-kasei.eu/investment-framework-for-carbon-neutrality/).

Libolon: Libolon company saves chemicals and water with fabrics produced from Ecoya® yarn and recycled polymer colored with natural pigments. In this way, it both protects the nature and demonstrates a good social approach (http://www.libolon.com/).

Hyosung Company: As an alternative to fibers produced from petroleum-derived raw materials, Hyosung company contributes to the creation of increasingly alternative options for the textile industry with the yarns it produces from sustainable materials. The manufacturer accelerates both more environmentally friendly and sustainable fiber production by making new investments in research and development to improve its fiber portfolio (https://sourcingjournal.com/sustainability/sustainability-materials/hyosung-synthetic-fiber-portfolio-creora-bio-based-spandex-recycled-nylon-471340/).

5 A Case Study: Polyteks Tekstil Fiber Production Company's Corporate Social Responsibility Studies

Founded in 1986 in Bursa Demirtaş Organized Industrial Zone (Bursa, Turkey), Polyteks Tekstil company is a polyester yarn manufacturer. Operating in the synthetic yarn sector, Polyteks continues its activities in an environmentally sensitive manner with its innovative approach. Polyteks, established on an area of 40,000 m^2, produces from very fine to thick yarns; it produces POY yarn, FDY yarn, textured yarn, and gimped yarn. With a manufacture capacity of 3600 t of yarn per month, the company serves many sectors, from home textiles to the automotive industry and from garments to carpets. In addition to its standard products, the company attaches a significant importance to R&D studies and specializes in functional yarns. A signatory of the Women's Empowerment Principles (WEPs), the company continues to work toward the 2030 Sustainable Development Goals. In this context, product life cycle analysis, carbon footprint calculation, and energy and waste management studies are among its activities. The company holds ISO 9001:2015, ISO 10002, ISO 14001, ISO 14001, ISO 45001, ISO 50001, OEKO TEX Standars100, and GRS certificates and carries out its activities within the framework of legal requirements and standards. It also has memberships such as CIRFS, SUSEB, TTGV, Techxtile Innovation League, and Women Empowerment Platform.

Aiming to meet the requirements of sustainability in every process from suppliers to customers, the company works to expand the use of bio-based and biodegradable raw materials as well as yarn production from recycled raw materials. In this context, it has also registered its own "POLY-bio" trademark. Adopting environmentally friendly approaches, the company contributes to the reduction of textile waste with innovative products such as dope-dyed yarn production and easy dyeable yarn production. With the projects it carries out, the company received the "One of the most innovative companies in Türkiye" and "Most sustainable project" awards at the TechXTile Start-Up Challenge event (https://tasdelengroup.com/tr/default).

Some of the yarns and threads produced as an outcome of the work implemented within Polyteks to minimize the damage caused by textiles to the ecology are as follows:

- Recycled yarns (Fig. 1)
- Polyester yarns and threads that can be dyed at low temperatures
- Dope-dyed yarns
- Biodegradable polyester yarns and threads
- Yarns and threads produced using bio-based raw materials

Growing day by day, Polyteks Tekstil fiber-producing company is making new investments and establishing a recycling plant, wastewater plant, and solar energy plant. Polyteks is investing in a recycling facility in order to obtain recycled raw materials by reducing plastic wastes, which constitute a serious burden on the environment, to reduce foreign dependency in this regard, and at the same time to prevent the formation of carbon footprint arising from procurement processes. With this facility, it is aimed to use the plastic waste generated in Turkey as a resource. Another investment is the wastewater plant. Since the use of chemicals and water is very high in textile processes, ecological approaches have started to be adopted. Thanks to the wastewater facility invested in this context, water savings will be achieved in the textile sector by reusing wastewater in the relevant industrial zone. At the same time, the company, which also works to reduce high energy consumption, will produce its own renewable energy with the solar energy facility investment and minimize its negative impact on the environment.

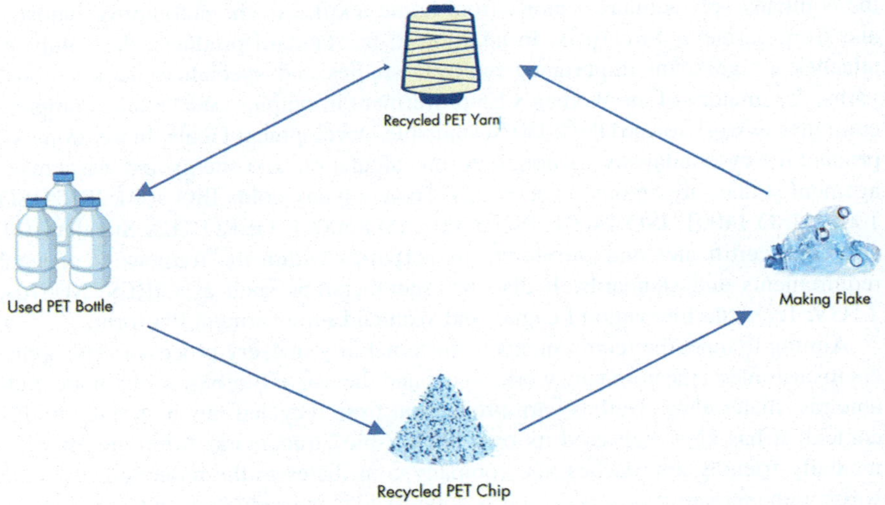

Fig. 1 Recycled yarn production cycle

Polyteks, which carries out many different activities within the scope of corporate social responsibility, cooperates with various NGOs and collects donations every year in the runs organized in Turkey by taking steps on behalf of non-governmental organizations such as Tohum Autism, Koruncuk Foundation, and Kaçuv. Every year, the company organizes events such as the April 23rd Children's Festival and the Republic Ball, as well as garbage collection events and social activities (Figs. 2 and 3).

Simultaneously, to foster a sense of belonging and benefit employees, donations are made on the birthdays of company employees, various awards are given depending on the time spent in the company, scholarship support is provided to the children of employees who are studying, and food is provided to employees throughout the year. The company supports organizations run by university students.

Polyteks, which is committed to sustainability and follows current studies, participates in many national and international organizations. In these events it follows the developments in the sector and meets with its stakeholders (Fig. 4).

Every year, Polyteks organizes internal training in many areas, such as combating violence against women, breast cancer, prostate cancer, and cervical cancer, in order to raise awareness among the company's employees (Fig. 5). These efforts mentioned within the scope of Polyteks company's corporate social responsibility are still continuing increasingly.

Fig. 2 Eker I Run 2023 Marathon

Fig. 3 Collection of waste plastic and garbage

Fig. 4 Sustainability Talks Organization

Fig. 5 International training on combating violence against women

6 Conclusion

Social responsibility is "the obligation of an institution or organization to maximize its positive effects on society and minimize its negative effects." Broadly speaking, it is "a concept that businesses should be actively concerned with the well-being of society." The social responsibility awareness of commercial organizations is rapidly increasing, and companies are starting to accept this concept. The textile industry is also included in this field. Companies that produce and trade textiles must also be aware of their responsibilities toward the society and the environment. These can be listed as follows, shareholders, government, customers, investors, suppliers, competitors, society, and environment. Corporate social responsibility is increasing in popularity today. Companies are increasingly changing their responsibilities to various other influences along with them. Companies know that they have to protect the interests of their employees, suppliers, and other vendors. For this reason, commercial companies that work only for the purpose of making profit may gradually be losing their existence in the long run. For this reason, many different textile fiber manufacturers, textile, and fashion companies producing in various countries of our planet are increasingly undertaking corporate social responsibility projects, studies, and new practices day by day.

References

Aktan, C. C., & Börü, D. (2007). Kurumsal sosyal sorumluluk. In *Kurumsal sosyal sorumluluk: İşletmeler ve sosyal sorumluluk* (pp. 11–36). IGIAD Publishing.

Aras, G., Tezcan, N., & Furtuna, Ö. K. (2018). Çok boyutlu kurumsal sürdürülebilirlik yaklaşımı ile Türk bankacılık sektörünün değerlemesi: Kamu-Özel banka farklılaşması. *Ege Akademik Bakis, 18*(1), 47–61.

Atılgan, T., Kanat, S., & İlleez, A. (2012). The environmental effects of corporate social responsibility development of enterprises in textile and clothing sector. *Akdeniz Sanat Hakemli Dergi, 4*(8), 73–79.

Bassen, A., Jastram, S., & Meyer, K. (2005). Corporate social responsibility: Eine Begriffserläuterung. *Zeitschrift für Wirtschafts-und Unternehmensethik, 6*(2), 231–236.

Bayraktar, A., & Cömert, C. E. (2018). Effect of corporate cultural responsibility on a Company's financial performance and brand image: An example of a Turkish Oil Company Opet AS. In *Corporate social responsibility for valorization of cultural organizations* (pp. 186–224). IGI Global.

Bianchi, C., & Birtwistle, G. (2012). Consumer clothing disposal behaviour: A comparative study. *International Journal of Consumer Studies, 36*(3), 335–341.

Birleşmiş Milletler Dünya Çevre ve Kalkınma Komisyonu. (1991). Ortak Gelece ğimiz, Belkıs Çırakçı (çev.). Türkiye Çevre Sorunları Vakfı Yayını, Ankara.

Boström, M., & Micheletti, M. (2016). Introducing the sustainability challenge of textiles and clothing. *Journal of Consumer Policy, 39*(4), 367–375.

Bozlağan, R. (2005). Sürdürülebilir gelişme düşüncesinin tarihsel arka plani. *Journal of Social Policy Conferences, 50*, 1011–1028. Istanbul University.

Burchell, J., & Cook, J. (2011). Banging on open doors? Stakeholder dialogue and the challenge of business engagement for UK NGOs. *Environmental Politics, 20*(6), 918–937. https://doi.org/10.1080/09644016.2011.617176

Can, Ö., & Ayvaz, K. M. (2017). Tekstil ve Modada Sürdürülebilirlik. *Akademia Sosyal Bilimler Dergisi, 3*(1), 110–119.

Carroll, A. B. (1991). The pyramid of corporate social responsibility: Toward the moral management of organizational stakeholders. *Business Horizons, 34*(4), 39–48. https://doi.org/10.1016/0007-6813(91)90005-G

Castka, P., & Balzarova, M. A. (2008). ISO 26000 and supply chains – On the diffusion of the social responsibility standard. *International Journal of Production Economics, 111*(2), 274–286.

Chan, R. Y. K. (2001). Determinants of Chinese consumers' green purchase behavior. *Psychology & Marketing, 18*(4), 389–413.

Chapin, F. S., III, Torn, M. S., & Tateno, M. (1996). Principles of ecosystem sustainability. *American Naturalist, 148*(6), 1016–1037.

Crane, A., Matten, D., & Moon, J. (2008). The emergence of corporate citizenship: Historical development and alternative perspectives. In *Handbook of research on global corporate citizenship* (pp. 25–49). Edward Elgar Publishing Ltd.

D'Orio, G., & Lombardo, R. (2007). Corporate governance and corporate social responsibility in Italy: Advantages and disadvantages of a non-explicitly-existent system. *Corporate Ownership and Control, 4*(4), 36–49.

Del Baldo, M. (2015). A state of the art of corporate social responsibility diffusion in Italy: Limits and potentials. In *Corporate social responsibility in Europe* (pp. 435–468). Springer.

den Hond, F., de Bakker, F. G. A., & Neergaard, P. (2007). Introduction. Managing corporate social responsibility in action: Talking, doing and measuring. In F. den Hond, F. G. A. de Bakker, & P. Neergaard (Eds.), *Managing corporate social responsibility in action: Talking, doing, measuring* (pp. 1–12). Ashgate.

Doğan, N. (2007). *İş etiğinin kurumsal sosyal sorumluluktaki rolü ve bir araştırma* (Doctoral dissertation). Marmara Universitesi, Türkiye.

Durgun, G., & Sarıbay, B. (2022). Sürdürülebilir bir dünya için ürün pazarlama stratejileri: Moda sektöründen örnekler. *Journal of Social and Humanities Sciences Research, 9*(81), 397–406. https://doi.org/10.26450/jshsr.2989

Engin, E., & Akgöz, B. (2013). Sürdürülebilir kalkınma ve kurumsal sürdürülebilirlik çerçevesinde kurumsal sosyal sorumluluk kavramının değerlendirilmesi. *Selçuk İletişim, 8*(1), 85–94.

Gedik, Y. (2020). kurumsal sosyal sorumluluk: tanımları, tarihi, teorileri, Boyutları ve Avantajları Üzerine kuramsal Bir Çerçeve. *Haliç Üniversitesi Sosyal Bilimler Dergisi, 3*(2), 265–304.

Geethamani, S. (2017). Advantages and disadvantages of corporate social responsibility. *International Journal of Applied Research, 3*(3), 372–374.

Goodpaster, K. E. (1991). Business ethics and stakeholder analysis. *Business Ethics Quarterly, 1*, 53–73.

Gürcüm, B., & Yüksel, C. (2012). Moda Sektörünü "Yavaşlatan" Eğilim: Eko Moda ve Moda'da Sürdürülebilirlik. In *1. Uluslararası Moda ve Tekstil Tasarımı Sempozyumu* (pp. 48–51). Akdeniz Üniversitesi Güzel Sanatlar Fakültesi.

Hoştut, S., & van het Hof, S. D. (2020). Greenhouse gas emissions disclosure: Comparing headquarters and local subsidiaries. *Social Responsibility Journal, 16*(6), 899–915.

http://www.libolon.com/. Accessed 5 Dec 2023.

http://www.unglobalcompact.org/docs/networks_around_world_doc/Network_material/CSR_Report_in_English.pdf. Accessed 10 Dec 2023.

https://about.nike.com/en/impact/focus-areas/protecting-the-planet. Accessed 19 Nov 2023.

https://about.nike.com/en/newsroom/reports/fy22-nike-inc-impact-report. Accessed 26 Oct 2023.

https://antex.net/sustainable-innovation/. Accessed 14 Oct 2023.

https://asahi-kasei.eu/investment-framework-for-carbon-neutrality/. Accessed 8 Dec 2023.

https://naia.eastman.com/. Accessed 8 Nov 2023.

https://sa-intl.org/programs/sa8000/. Accessed 15 Oct 2023.

https://sourcingjournal.com/sustainability/sustainability-materials/hyosung-synthetic-fiber-portfolio-creora-bio-based-spandex-recycled-nylon-471340/. Accessed 4 Dec 2023.

https://tasdelengroup.com/tr/default. Accessed 2 Dec 2023.

https://www.dowaksa.com/sustainability.html. Accessed 18 Nov 2023.

https://www.dyneema.com/about-us/latest-news/blog/dyneema-provides-innovative-and-sustainable-solutions-for-fabric-and-protective-knitwear. Accessed 27 Dec 2023.

https://www.globalcompactturkiye.org/10-ilke/. Accessed 12 Oct 2023.

https://www.huvis.com/mobile/eng/company/GreenBusiness.asp. Accessed 9 Nov 2023.

https://www.inditex.com/itxcomweb/en/sustainability. Accessed 25 Oct 2023.

https://www.indoramaventures.com/en/sustainability/social/csr. Accessed 14 Dec 2023.

https://www.indoramaventures.com/en/sustainability/social/csr#recycling-education. Accessed 22 Nov 2023.

https://www.invista.com/stewardship/social. Accessed 24 Nov 2023.

https://www.iso.org/obp/ui/en/#iso:std:iso:26000:ed-1:v1:en. Accessed 13 Oct 2023.

https://www.kolonindustries.com/en/sustainability/esg/report. Accessed 17 Oct 2023.

https://www.kuraray.com/csr. Accessed 11 Nov 2023.

https://www.lenzing.com/newsroom/press-releases/press-release/lenzing-awarded-platinum-status-for-sustainability-by-ecovadis. Accessed 25 Oct 2023.

https://www.mavi.com/sustainability/index.html. Accessed 23 Nov 2023.

https://www.mavicompany.com/tr/degerler/sosyal-sorumluluk. Accessed 25 Nov 2023.

https://www.m-chemical.co.jp/en/csr/pdf/csr_report_mrc_pdf. Accessed 10 Oct 2023.

https://www.natureworksllc.com/sustainability/sustainable-feedstocks. Accessed 27 Oct 2023.

https://www.nike.com/tr/surdurulebilirlik/hizmetler. Accessed 25 Oct 2023.

https://www.nilit.com/nilit-enriches-corporate-social-responsibility-programs-and-publishes-formal-sustainability-business-strategy-based-on-life-cycle-assessment/. Accessed 15 Nov 2023.

https://www.owenscorning.com/en-us/corporate/sustainability/docs/2023/2022-Owens-Corning-Sustainability-Report.pdf. Accessed 20 Oct 2023.

https://www.radicigroup.com/en/news-media/news/100-naturally-sourced-yarn-radicigroup-launches-biofeel-eleven-72512. Accessed 3 Dec 2023.

https://www.sasa.com.tr/surdurulebilirlik. Accessed 16 Oct 2023.

https://www.sasa.com.tr/surdurulebilirlik/sosyal-surdurulebilirlik/sosyal-sorumluluk. Accessed 17 Nov 2023.

https://www.toyobo-global.com/pdf/sustainability/report/2022/sustainability2022_e.pdf. Accessed 12 Nov 2023.

https://www2.hm.com/tr_tr/sustainability-at-hm.html. Accessed 16 Oct 2023.

İstanbul Menkul Kıymetler Borsası. (2011). *İMKB İstatistik Müdürlüğü*. Sürdürülebilirlikle İlgili Özet Bilgiler.

Jones, B., Bowd, R., & Tench, R. (2009). Corporate irresponsibility and corporate social responsibility: Competing realities. *Social Responsibility Journal, 5*(3), 300–310.

Kotler, P., & Lee, N. (2016). *Kurumsal sosyal sorumluluk*. İstanbul MediaCat Kitapları.

Krasodomska, J. (2013). Corporate social responsibility as a factor influencing the development of social accounting and assessment of employers. *e-Finanse: Financial Internet Quarterly, 9*(1), 12–25. ISSN 1734-039X.

Lai, C. S., Chiu, C. J., Yang, C. F., & Pai, D. C. (2010). The effects of corporate social responsibility on brand performance: The mediating effect of industrial brand equity and corporate reputation. *Journal of Business Ethics, 95*, 457–469.

Lindgreen, A., & Swaen, V. (2010). Corporate social responsibility. *International Journal of Management Reviews, 12*(1), 1–7.

Macit, N. (2020). *BİST Şirketlerinin Kurumsal Sosyal Sorumluluk Faaliyetleri: BİST 30, BİST 50, BİST 100 Endeksleri Üzerine Karşılaştırmalı Bir İçerik Analizi*. Yüksek Lisans Tezi, Karamanoğlu Mehmetbey Üniversitesi Sosyal Bilimler Enstitüsü.

Maignan, I., & Ralston, D. (2002). Corporate social responsibility in Europe and the U.S.: Insights from businesses' self-presentations. *Journal of International Business Studies, 33*, 497–514.

McKinsey & Company. (2022). State of fashion 2022: An uneven recovery and new frontiers. https://www.mckinsey.com/industries/retail/our-insights/state-of-fashion

McWilliams, A., & Siegel, D. (2001). Corporate social responsibility: A theory of the firm perspective. *Academy of Management Review, 26*(1), 117–127. https://doi.org/10.2307/259398

Metlioğlu, H. H., & Yakin, V. (2021). Tekstilde sürdürülebilirlik: Hızlı moda markalarının sürdürülebilirlik stratejileri. *OPUS International Journal of Society Researches, 18*(Yönetim ve Organizasyon Özel Sayısı), 1883–1908.

Montabon, F., Melnyk, S. A., Sroufe, R., & Calantone, R. J. (2000). ISO 14000: Assessing its perceived impact on corporate performance. *Journal of Supply Chain Management, 36*(1), 4–16.

Nalbant, Z. E. (2005). İşletmelerde Sosyal Sorumluluk ve İş Ahlakı. *Yönetim ve Ekonomi, 12*(1), 197.

Özcan, G. (2021). *Yeni Tüketici Açısından Kurumsal Sosyal Sorumluluk Yaklaşımının Kurumsal itibar ve Tüketicilerin Tavsiye Etme Niyetine Etkisi Üzerine Bir Araştırma* (Doctoral dissertation). Marmara Universitesi, Türkiye.

Sachs, J. D., Lafortune, G., Fuller, G., & Drumm, E. (2023). *Sustainable development report 2023: Implementing the SDG stimulus*. Dublin University Press.

Seyitoğullari, O., & Bilen, A. (2020). Kurumsal Sosyal Sorumluluk Üzerine Bir Literatür Taraması. *Sosyal Bilimler Akademi Dergisi, 3*(2), 195–205.

Sweeney, L., & Coughlan, J. (2008). Do different industries report corporate social responsibility differently? An investigation through the lens of stakeholder theory. *Journal of Marketing Communications, 14*, 113–124.

Tandon, N., & Reddy, E. E. (2013). A study on emerging trends in textile industry in India. *AMET International Journal of Management, 5*(5), 81–88.

Tokgöz, N. (2018). Genel İşletme. In Z. Erdoğan & & A. Hepkul (Eds.), Eskişehir: T.C. Anadolu Üniversitesi Yayını No: 2650, Açıköğretim Fakültesi Yayını No: 1616. https://ets.anadolu.edu.tr/storage/nfs/ISL107U/ebook/ISL107U-12V4S1-8-0-1-SV1-ebook.pdf. Accessed 8 Oct 2023.

Tsoutsoura, M. (2004). *Corporate social responsibility and financial performance* (Working paper series, paper 7) (p. 3). Center for Responsible Business.

Turancı, E. (2021). Kurumsal Sosyal Sorumluluk Bağlamında Modada Geri Dönüşüm ve Sürdürülebilirlik: Göstergebilimsel Bir Analiz. *Erciyes İletişim Dergisi, 8*(1), 347–371. https://doi.org/10.17680/erciyesiletisim.801002

Turker, D. (2009). Measuring corporate social responsibility: A scale development study. *Journal of Business Ethics, 85*(4), 411–427. https://doi.org/10.1007/s10551-008-9780-6

Turpcu, Y. (2021). İşletme Misyonunun Sosyal Sorumluluk Açısından İncelenmesi: BİST 30 Endeksi Üzerine Bir İçerik Analizi. *Journal of Academic Value Studies, 7*(4), 469–491.

UİB. (2017). *Hazır Giyim Sektöründe Sürdürülebilir Trendler*. Uludağ İhracatçı Birlikleri Genel Sekreterliği.

Vermeir, I., & Verbeke, W. (2008). Sustainable food consumption among young adults in Belgium: Theory of planned behaviour and the role of confidence and values. *Ecological Economics, 64*, 542–553.

World Business Council for Sustainable Development. (2020). Corporate social responsibility: Making good business sense. http://www.ceads.org.ar/downloads/Making%20good%20business%20sense.pdf. Accessed 16 Oct 2023.

Yavuz, V. A. (2010). Sürdürülebilirlik kavramı ve işletmeler açısından sürdürülebilir üretim stratejileri/concept of sustainability and sustainable production strategies for business practices. *Mustafa Kemal Üniversitesi Sosyal Bilimler Enstitüsü Dergisi, 7*(14), 63–86.

"Corporate Social Responsibility in Textiles and Fashion": The Strategic Shift from *Circular* to *Insular* Economy

Alessia Agostini, Cassiopée Chastin, Lucie Gonzalez, and Ivan Coste-Maniere

Abstract This chapter delves into the transformative shift from a circular to an impressive insular economy, signifying a paradigm change that transcends mere economic and environmental thinking. Recent years have discussions surrounding the circular economy, driven by pressing crises like the COVID-19 pandemic and climate change. These crises compel us to undertake a profound reassessment of our perspectives, beliefs, and practices. Circular economy is a new academic powerful keyword and hot topic.

Keywords Insular economy · Circular economy · Greenwashing · Sustainability · Climate change · Natural fibers · Textiles · Fashion · Long-term sustainable advantage

1 Introduction

The imperative acknowledgment that the circular economy, once hailed as the ultimate solution for sustainability, frequently falls short of genuine circularity. It reveals inherent limitations in the pursuit of sustainability. Disturbingly, certain sectors, such as the textile industry, have instrumentalized the concept of the circular economy for greenwashing, eroding its authenticity and credibility. This sobering reality urges us to critically assess the pertinence and authentic utility of the circular economy.

Considering these concerns, we are compelled to question whether the concept of the circular economy remains pertinent or has devolved into a superficial response to the pressing sustainability challenges of our era. Does it still serve as a meaningful framework for addressing the multifaceted aspects of sustainable development,

A. Agostini · C. Chastin · L. Gonzalez · I. Coste-Maniere (✉)
MSc Luxury & Fashion, SKEMA Business School, Lille, France
e-mail: ivan.costemaniere@skema.edu

S. S. Muthu (ed.), *Corporate Social Responsibility in Textiles and Fashion*,
Sustainable Textiles: Production, Processing, Manufacturing & Chemistry,
https://doi.org/10.1007/978-3-031-61099-8_11

or has it become a convenient façade to address immediate concerns? Fast communication in this field should be avoided.

The insular economy represents a profound and potentially much more comprehensive solution. It spotlights the paramount significance of sustainability and regenerative practices as it pertains to global sustainability and sustainable processes and pathways. This prompts us to ask: What valuable lessons can we extract from the insular economies of regions such as Mauritius, Maldives, Oman, Seychelles, and Zanzibar, where sustainability and resilience are deeply ingrained? By studying these insular paradigms, we aim to uncover sustainable processes and best practices that could serve as catalysts for global sustainability.

Throughout this chapter, we will draw insights from the existing case studies, such as initiatives involving upcycling marine plastic debris and fishing gear from island coastlines, including locations like the Maldives and Sri Lanka, into high-performance products. Moreover, innovative companies understood the need to return to natural raw materials, including bamboo, tapa, linen, and coco fibers. Or even oyster wastes or abalone shells and floated woods…. The Italian company Orange Fiber has perfectly responded to this demand, harnessing waste from the food industry as a natural raw material for textile production. These examples serve as inspirations for sustainable practices and innovative approaches within the insular economy paradigm.

In these insular regions, sustainability is not a passing trend, but an enduring imperative deeply rooted in their cultural ethos. Amidst their distinctive contexts, vulnerability appears as a stark reality due to limited resources, both natural and economic. Yet, in the face of these resource constraints, their response illustrates an unwavering long-term commitment and a dedication to lasting solutions. They offer profound lessons in reconstructive sustainability and autonomous regeneration, seeing social responsibility as an integral part of their identity rather than a corporate obligation.

These findings carry significant implications for industries globally as they navigate the path toward a more sustainable and responsible future. Furthermore, this new line of thinking and findings compels us to reconsider the question of what it truly means to be sustainable. Don't you think, the most sustainable people are the ones who have learned to thrive with the least?

To address these questions, we employ a multifaceted approach to investigate the potential of insular economies as a leading model of sustainability and their implications for the luxury and fashion industries. Our chapter is divided into three distinct sections.

The chapter's first section explores the rise and fall of the circular economy, initially hailed as a promising sustainable alternative to the linear economy. Not only a semantic shift to face as a major issue…Through a combination of academic review and real-world case studies, it examines the concept's shortcomings, including its vulnerability and exposure to greenwashing and the ineffectiveness of its implementation for numerous industries. Recognizing the need for a new approach, the section introduces insular economies as a potential solution.

The second section begins with an in-depth examination of the concept of insular economies, highlighting their inherent resilience and ingenuity as a model of sustainability. To provide an overview, we use descriptive methods to examine their sustainability practices. Next, we examine three case studies—La obra de llatra in Mallorca, Spain, Casa Flor Ixcaco in Guatemala, and the “slow fashion” movement—to illustrate the practical applications of insular economy principles. Finally, we conclude this section by summarizing the key takeaways from the case studies and emphasizing the potential of insular economies to serve as models for a more sustainable future.

Finally, we end our chapter with a section exploring how insular economies can revolutionize the luxury and fashion industries through sustainable materials and ecotourism. We begin by examining reconstructive sustainability, a concept that focuses on the creation of new materials from renewable resources to meet the growing demand for eco-friendly products. Beyond material innovation, we explore the synergistic relationship between insular economies and ecotourism, showing how sustainable luxury resorts can benefit island communities and the environment.

1.1 The Fall of a Highly Praised New Model: The Circular Economy

1.1.1 The High Hopes and Promises Brought by the Circular Economy

In a world in climate crisis, where climate hazards are happening more and more often and are more and more destructive, where species are more and more endangered, NGOs and civil society grow awareness toward the climate change and the risk of biodiversity loss. Industries and companies—as contributing to raise the greenhouse gas emissions—are often designated as the main polluters leading to increase the climate change. Now, civil societies and consumers expect from economic actors to take the lead on the environmental and social issues and to instill a real change, to stop being catalysts of the change.

For a couple of years, a new economic paradigm has been arising: the circular economy. This model appears to be a way to instill the change needed and required by civilians. In fact, the circular economy has highly praised goals, such as reducing greenhouse gas emissions, pollution, and waste to slow down the global warming and avoiding the excessive consumption of natural resources and rather keep them in a close circle to reuse them and regenerate nature (Ellen MacArthur Foundation, n.d.).

The circular economy wishes to replace the linear economy of *Take, Make, Waste* (Cheng, 2021) by an economy of *Reduce, Reuse, Recycle* (DiSalvo, 2019). This shift has been mainly pushed by the new expectations of customers, especially from the younger generations (Millennials and Gen Z). Indeed, these latter have developed an important concern toward the environment and the conditions of production (social, environmental, and health). To match the new expectations of consumers, companies had to adapt their policies and governance. They must tackle environmental and social issues in their ruling and production processes. To do so, companies must go beyond the law, to do more than the law is requiring as there is—at the moment—a governmental void on these issues (Centobelli et al., 2022). There is a start of regulation in the fashion industry with the push of Europe with its *Strategy for Sustainable and Circular Textile Development* (Centobelli et al., 2022). Its aim is by 2030 all textiles sold in the EU will be safe for health, ethically produced, respecting the environmental guidelines and people, long-lasting and recyclable, and mainly made of recycled materials (Centobelli et al., 2022). It is great that some regions take the lead on these topics, but it seems to be too basic and not ambitious enough. These are really suited to the mere bases of the circular economy and relying on common sense and on humanity.

1.1.2 A Model That Loses Its Luster and Has Been Derived Toward Greenwashing…

In its early ages, the circular economy has been considered as bringing a revolutionary change in the economy industry, pushing companies to think about the whole process, and life cycle of their products. The main aim was to reduce waste, energy consumption, and surplus. Firms have taken the lead and gone beyond the mere prerequisites of law. So, in various sectors, for quite a few years, companies have already made some change regarding the conception of their products (e.g., made of recycled materials) or in their end of life (being recyclable). But sadly, most of them turned to the easiest solution: make a slight change in the product elaboration to be able to say it comes from recycled fabrics (reuse pillar) or by making it recyclable (recycle pillar). As a result, brands didn't really address and fit to the circular economy as most of them fear to suffer from a loss of profits. Firms don't take the time to rethink their whole product cycle and the design of these products, so they mostly end up doing greenwashing, even if it would have been more effective on the long run to redesign the whole production process of products (Cabanes & Cruaud, 2023).

It can be observed that through the spreading of this circular theory, its meaning has been diluted, being too often reduced to using recycled materials—or making their products recyclable—without thinking about the required surplus of energy. These are the easiest and cheapest, short-term solutions adopted by the wide majority of companies, and these solutions are more belonging to the recycling economy (Hepster, 2023) than to a real circular economy. For this reason, many critics accuse the circular economy to just be a tool of greenwashing and to not bring any concrete change toward a greener economy.

The *Adidas Ocean Plastic* sneaker is a good example of the use of circular economy as greenwashing. This model of sneakers is made from big chunk of plastic collected on beaches (and not in oceans as the name suggests it). At first glance, this appears to be a good initiative, but it distracts from the problem. The issue is not the reusability of plastic fund in natural environments, but the real problem to tackle is the possibility of plastic to reach shores and oceans (Domingo, 2021). Moreover, bringing the collected chunk of plastic from the coast to the factory is more polluting than the reuse of local plastic (Domingo, 2021). Thus, this pair of sneakers isn't really a part of circular economy, and it could be more considered as a simple act of greenwashing.

This initiative shows the lack of sustainability and of real changes. The circular economy has mostly contributed to increase the cases of greenwashing. Companies turn to easy, quick, short-term, or fake answers. They are really limited and failed most of the time as they are not enough well thought or eco-conceived. The change led by firms seems to have had a slight or negligeable effect and can be more linked to greenwashing than a real eco-conscious change.

The garment and fashion industry is no exception in the circular economy, mostly being greenwashed. Indeed, now fast-fashion brands are offering a whole new segment of clothes made of recyclable materials. In doing so, they want to appear greener and to comply with the circular economy paradigm. But it is a scam, and they are not addressing to the core of the problem, which is their business model—relying on overproduction. These brands cannot by essence belong to the circular economy.

Should we be considering as astonishing fast-fashion industry is contributing to the deterioration of the environment? Eighty-five percent of the clothes produced by the fast-fashion brands end up in landfills or are incinerated (Yeoh, 2018). Every second a landfill receives the equivalent of a garbage truck full of clothes (Igini, 2022, n.p.). The incineration of these bad-quality made-of-plastic clothes releases a big amount of carbon emission. The fashion industry contributes between 2% and 8% of global carbon emissions (Dontoh & Pandya, 2022). Furthermore, in 48 h, H&M sells a quantity of clothing that would take up to 12 years to recycle (Agro, 2018). All these facts show the fast-fashion industry cannot by essence integrate circular economy solutions without totally rethinking and changing their business model. It would first start with a drastic reduction of their collections' size and frequency.

On another note, the fast-fashion clothes are hardly recyclable as they are of a too poor quality (Agro, 2018) and made of a heterogeneous mixture of synthetic materials (Coulson-Drasner, 2021) which are almost impossible to split. As a result, less than 1% of used clothes are recycled into new ones (Dontoh & Pandya, 2022). Most of fast-fashion clothes are made of polyester which derives from petrol and requires a lot of energy to do—releasing carbon emission during the process (Coulson-Drasner, 2021). Moreover, during each wash, the garment made of polyester releases thousands of micro-plastics into water, which will end up in rivers and later in our food (Coulson-Drasner, 2021).

So, for now, the reducing and recycling pillars are not valid in the fast-fashion industry. What about the reuse pillar? Since 2010, the time clothes are worn has considerably decreased, and they are worn approximately 36% less than before (Ellen MacArthur cited in Dontoh & Pandya, 2022). They end up being donated or thrown away. But, as they are of a too poor quality to be resold or reused in Western countries, they are sold and shipped to African countries (e.g., Ghana) (Insider

Business, 2022). There, the local population receives impressive amount of old fast-fashion clothes, which are displayed at the Kantamanto market (Insider Business, 2022). The clothes are of a too poor quality, often stained or ripped, so they are hard to sell. As a result, more than a third of them directly end up in landfills. Their landfills can reach 30 meters high and are contributing to flooding and spreading of diseases, like malaria (Insider Business, 2022). Moreover, garments pollute the natural environment, can travel over several kilometers, and can reach beaches that are close by (Insider Business, 2022).

For the fast-fashion industry, the recycling and reusing (upcycling) pillars must be handled very carefully and be addressed in a more effective way. The reuse and recycle patterns followed by brands are not enough and not well thought enough; in fact greenhouse gas emissions have not been reduced by recycling everyday items (Domingo, 2023). One should be aware of the environmental cost of the transport, and manufacturing recycled products can have a more detrimental impact than burning plastics closed to where they have been used (Domingo, 2023).

1.1.3 The Irrepressible Urge for a Needed Shift…

All these facts, datas, and examples showcase the need for a new shift in paradigm. The circular economy did not reach its initials goals and has been diverted toward greenwashing. Currently, the promises brought by this circular paradigm seem to be already an impasse even if some companies just stared to apply circular principles.

The economic actors must now be in quest of a solution which will bring a radical change in the practices and running of companies. A new insight should be taught by insular civilizations, who have no other choice to respect the environment and the local biodiversity and to reuse everything as they have limited and very scarce resources available. This scarcity can lead firms to rethink everything throughout their processes, to make the best out of each material or of the energy consumed. Each step and each process is well thought out before starting the making, to be sure there is no overproduction, no loss of energy, no pollution, and no harm for the environment, the population, and the biodiversity during the fabrication, the consumption, or the end life of the product.

1.2 The Rise of a New Paradigm: Back to the Roots

1.2.1 In Search of Resilience: Insular Economy as a New Intrinsic Powerful Model for the Environment

When we are facing challenges, doubt, and difficulties, isn't it wise to return to our roots? On the one hand, revisiting our cultural and family origins, reconnecting with our fundamental principles, and reflecting on who we once were are not signs of weakness. On the contrary, it's a testament to intelligence and pragmatism. On the other hand, we have sometimes to remember how we thrived in times of scarcity, as it was often in those moments that our resourcefulness shone brightest.

This is what we also observe in island countries, through their insularity, stand as a compelling, a great example of using their ancestral knowledge and making the most of limited resources. Their insularity refers to a condition characterized by a space or territory closely associated with the concept of an "island," where isolation may result either from inherent geographic isolation or from a deliberate act of protectionism or self-isolation. This unique situation allows them to think and act differently.

Unfortunately, island nations are frequently perceived as frail, vulnerable, and, to some, detrimental to the environment due to their geographic isolation. Being physically separated from larger landmasses restricts their resource access and results in significant dependence on imports. Additionally, their small size makes achieving economies of scale in various industries a daunting task. The limited land area and resources also impact their ability to sustain growing populations and diverse economic activities. Furthermore, isolation can lead to a lack of connectivity and infrastructure.

However, these characteristics, often seen as flaws, can in fact be transformed into strengths, paving the way for a sustainable future. It could be said that *less is more*: in these apparent weaknesses, we discover our greatest assets. The constraints associated with the "island" concept, whether due to geographical isolation or a deliberate choice to remain isolated, create a unique situation. Their limited access to many resources enables them to minimize waste, streamline processes, and reduce pollution. This exceptional resilience makes the 47 island nations, out of the 193 member states of the United Nations, ideal models and solutions for our future efforts.

These island states seek autonomy and, despite their resources limitations, have developed strategies to thrive within the confines of their ecosystems. Their low-input, resilient technologies maximize the use of natural processes. In a world faced with environmental degradation and resource depletion, these lessons become invaluable.

The insular economy, deeply rooted in the cultural ethos of insular regions, emphasizes sustainability and regenerative practices. Facing resource constraints, island nations demonstrate an unwavering, long-term commitment to lasting solutions. They offer profound lessons in reconstructive sustainability and autonomous regeneration, seeing social responsibility as an intrinsic endogenic integral part of their identity rather than a corporate obligation.

The challenges of our time are forcing us to rethink our approach to sustainability, economics, and resource management. The insular economy offers a new path, born of constraints, vulnerability, and the instinct to return to resources. It challenges us to see strength where we once saw weakness and to consider the value of resourcefulness in a world of abundance.

To better understand how these concepts work in practice, let's look at a few case studies.

1.2.2 Examples of Successful Insular Economies

The *Obra de Llatra* or the Art of Mallorcan Rattan

The *obra de llatra*, also known as *obra de palma*, is a unique artisanal tradition in Majorca, an island in the Balearic Islands of Spain. It revolves around the fan palm tree, *Chamaerops humilis*, the only native palm tree in Europe and the Mediterranean. The leaves of this palm tree are manually harvested, primarily in the mountainous areas of Andratx, Pollença, Artà, and Capdepera, starting from the month of July. The production process of obra de llatra is a demanding art that begins with sun-drying the leaves on shelves for approximately 20 days. Once dried, the leaves are carefully separated from the palm, and the most beautiful and whitest strands are reserved for crafting fine articles, while the remaining strands are used for making brooms and brushes, resulting in minimal waste.

The artisanal process continues with immersing the palm leaves in water and soap, which bleaches them and makes them more pliable for crafting artisanal objects. Additionally, the strands can be dyed using natural products like plants, depending on the intended use, providing a wide range of colors for the finished products. Majorcan artisans skillfully use their hands to create a variety of items, including baskets (referred to as *cenachos*), hats, bags, brooms, brushes, and much more. These products are esteemed for their quality, durability, uniqueness, and eco-friendly approach.

The obra de llatra plays a crucial economic role in Majorca, as it generates local employment and contributes to the tourism industry by offering unique artisanal products to visitors. This serves as an example of an island economy that values local natural resources, maintains a cherished artisanal tradition, and avoids environmentally harmful processes. Majorca's rattan craftsmanship aligns with the principles of corporate social responsibility (CSR) by promoting sustainable practices, preserving the region's biodiversity, and providing a livelihood for local artisans, thereby contributing to the economic stability of the island community.

Casa Flor Ixcaco: Maya Indigenous Weaving Styles in Guatemala

Guatemala is a continental country without any islands. However, certain mountainous and rural regions of the country can be quite remote and isolated, largely because of the challenging and rugged terrain. In these isolated areas, communities have established local economies that are deeply rooted in their traditions, skills, and the resources that are accessible to them. This unique economic setup can sometimes resemble an island economy.

The textile tradition in Guatemala holds a great importance and deeply rooted Mayan culture. The weaving techniques, vibrant patterns, and meaningful colors all of which convey profound cultural and symbolic messages that reflect the history and spiritual beliefs of indigenous communities.

Casa Flor Ixcaco is an example of a cooperative that preserves indigenous Mayan weaving traditions while addressing environmental concerns and corporate social responsibility (CSR) in its business model. Situated in San Juan la Laguna, nestled along the shores of Lake Atitlán in Guatemala, this cooperative exemplifies how its artisans create their distinctive fabrics.

In the selection of raw materials, Casa Flor Ixcaco demonstrates its commitment to sustainability by growing its own cotton in four native varieties of different colors (white, khaki, ixcaco, and green), all grown in the mountains of their region. This conscientious approach not only safeguards local cotton diversity but also diminishes the environmental impact associated with raw material transportation. Furthermore, it's worth noting that the cotton they use is 100% organic.

The process of cotton preparation and spinning begins with raw cotton being beaten with pieces of wood to prepare it for spinning. Skilled weavers employ traditional spindles to transform the cotton into thread. The spindle is a stick-shaped tool with a bobbin around which the thread is wound. This spinning technique is not only vital for producing high-quality yarn but also plays a pivotal role in preserving the rich tapestry of Mayan culture and artisanal skills.

Dyeing the yarn is an important step. The cooperative uses natural dyes derived from plants, fruits, seeds, and flowers to color the yarns. These natural pigments are sourced by local plants, and traditional dyeing methods are used. For instance, shades of green are achieved using cinnamon, while curry produces yellow-orange, and beet results in bright pink. This environmentally conscious approach relies on local flora to provide vibrant colors, eliminating the chemical dyes that are potentially polluting and dangerous for humans.

Weaving: Once the threads have air-dried, they are skillfully woven using a wooden hand loom, which can be set up in various locations, sometimes hung from a tree or a sturdy post. Hand weaving allows for precise control of thread tension and intricate pattern creation, resulting in truly unique pieces.

Each creation at Casa Flor Ixcaco represents more than just a piece of fabric; it reflects a life, a culture, and a community. A deep respect for every stage of the process, from material selection to the artistic creation, underscores the integral role of corporate social responsibility (CSR) in their approach. Every fabric, often requiring a month's labor, is the result of a responsible mode of production that honors both the environment and culture. It embodies the essence of an economy rooted in the local availability of resources. As one of the women working at Casa Flor Ixcaco aptly expressed, "It's not just a piece of fabric; it's a reflection of the life behind each creation."

1.2.3 Revolutionizing Fashion: The Artisanal and Local Approach in Insular Economies

In today's world, as we face the alarming consequences of our relentless overconsumption, the harsh reality of environmental degradation, and the urgent need for sustainability, the fashion industry emerges as a prominent player in these unfolding challenges. The rise of fast fashion has led to an era of excessive waste, worker exploitation, and environmental degradation, exacerbating the global crisis of climate change.

In recent years, however, there has been a growing awareness of the need for sustainability in fashion. But for the time being, this is only a façade, as recycling and the use of eco-friendly materials often rely on energy-intensive machines or manufacturing processes, contributing to pollution and other environmental problems. These awareness and change have started to influence consumer behavior and will keep evolving. By 2030, consumers are expected to distance themselves from fast-paced lifestyles and excessive consumption. They are increasingly embracing a slower, more minimalist approach to consumption, giving higher importance to durability, protection, and functionality. This shift reflects a growing demand for consumers to buy less and choose quality and long-lasting products, making "slowness" an ideal state and moving away from desire for speed.

In the case of the above-mentioned insular economy, the main focus is on craftsmanship, local production, and the utilization of natural resources and materials. These serve as prime examples of how the fashion and textile industry can be redefined and rethought to minimize its environmental footprint. This approach aligns closely with the principles of the "slow fashion" model, where product quality and longevity take the center stage, waste is minimized, and positive social impact is maximized. Therefore, artisanal practices serve as a model for both slow fashion and a sustainable future.

Choosing local products and supporting artisans are significant steps toward adopting a more responsible and environmentally friendly lifestyle. This decision not only benefits us but also contributes to a better future. Artisanal products prioritize quality over quantity, and they are marked by meticulous attention to detail, expert craftsmanship, and the use of high-quality materials. Artisans employ traditional techniques that have been honed over centuries, resulting in hand-crafted garments that are exceptionally durable, reducing the need for frequent replacements. These enduring creations are a testament to the artisans' dedication and skill. Moreover, this choice supports the local economy, fostering diversity within the textile industry. It also plays a crucial role in preserving traditions and expertise that are gradually being displaced by automated machinery.

The significance of craftsmanship in today's context arises from its role in addressing the adverse consequences of our excessive consumption and our commitment to safeguarding the environment. It extends beyond being a mere choice; craftsmanship represents a deliberate commitment to sustainability and environmental preservation. This commitment is founded on core principles that translate into a positive influence on our planet. It is synonymous with local and short-circuit rationalized production. Handcrafted items are often created in proximity to their consumers, as illustrated in the examples mentioned earlier, promoting local economies and concurrently diminishing the carbon footprint associated, for example, with transportation. This approach also diminishes reliance on expansive global supply chains, thereby fostering greater economic resilience.

Craftsmanship places sustainability and environmental respect at its core. Handcrafted products are meticulously designed to minimize their environmental footprint, contributing to the preservation of our planet. They lie in its efficiency in resource utilization, resulting in less waste. This approach significantly reduces production waste, a pivotal factor in preserving our planet. Craftsmanship stands as a long-term solution for both our planet and society. By prioritizing quality, authenticity, and sustainability, it plays a pivotal role in reshaping our consumption patterns. This transformation fosters a world where environmental values take the center stage, creating a positive and enduring impact.

This choice transforms our purchasing habits and the emotions associated with acquiring a product. When we invest in a handcrafted item, we are investing in a unique piece that carries a story. This nurtures a sense of compassion for what we wear and the communities behind its creation. We feel a heightened sense of pride in our purchases, knowing precisely where they come from, what materials were used, and creating an authentic bond with the artisans.

However, it is essential to ask ourselves a crucial question: whereas craftsmanship was once the norm for our ancestors, being the only way to acquire clothing, food, and essential objects, today it is often perceived as a luxury product. We can legitimately wonder about the possibility of making crafts accessible to all in our future. Can it once again become a mainstream option rather than the preserve of an elite?

1.3 The Insular Economy and Its Possible Translation in the Luxury and Fashion World

In recent decades, the concept of sustainability has become a prominent concern in the public debate and political agendas. This concern comes from the complex environmental crisis we are facing at a planetary scale, especially due to high-polluting and resource-intensive sectors. Among luxury segments, fashion and textiles are the second-highest contributor to carbon emissions, after the oil sector (SUN, 2023). Indeed, every year the fashion industry generates a conspicuous impact on the environment; it accounts for up to 10% of global carbon dioxide output and for a fifth of the 300 million tons of plastic produced globally each year (Bloomberg, 2022).

In this context, circular economy is expected to address this issue, bringing innovation along all the product life cycle and changing the traditional linear economy, which is based on *take-make-dispose* model. However, there is still a big discrepancy between theory and practice. In fact, most of the research on sustainable business models is still exploratory, and *sustainability* is still associated with vague and ambiguous notions. Even if this concept has been disseminated in the last decades but not concretely translated into practice, it is still the best direction to follow in the twenty-first century. Thus, it is imperative to close the gap and offer a framework that can go beyond theory and toward a *reconstructive sustainability*.

This objective should go through the awareness of the inherent ambiguity surrounding the term *circular economy*, leading the shift to the concept of *insular economy*, where sustainability is not a passing trend, but an enduring imperative deeply rooted in their cultural ethos. As discussed in the previous chapter, insular economies are key examples of regenerative systems, which focus on creating synergies that regenerate constantly the natural capital and services, through a *recover, restore, and regenerate* paradigm. Islands are examples of resilient and sustainable models, capable of exploiting the exposure to exogenous disturbances, the scarcity of resources, and the fragility of environmental systems and promoting the interaction of the main economic and environmental variables. One case is the Mauritius district, created within a free-export process zone and specialized in textiles, which has hugely evolved over 15 years, from peripheral low-wage industries to more technological ones, reaching an outstanding long-term growth. Mauritius has achieved a remarkable success in the textile and apparel industry, through a visionary approach, based on diversification, sustainability, and innovation.

The luxury and fashion sectors should learn from these insular realities, where permanent handicaps (including limited resources, higher transportation costs, vulnerability to external shocks) are transformed from weaknesses into assets, harnessing their unique advantages, such as distinctive natural resources, tourism potential, and strategic locations. The recovery of natural resources become an imperative in the current world, where the fashion and textile industries consume large amounts of water and release tons of toxic chemicals. Suffice it to say that about 40% of colorants used worldwide contain chlorine and other toxic dyes, a known carcinogen. Organic fibers could be the perfect samples of *reconstructive sustainability*: organic linen, hemp, milk silk, corn fiber, bamboo fiber, orange fiber, soy silk, pineapple, and banana fibers are the future of luxury and fashion industry (Muthu, 2017).

1.3.1 The New Frontiers of Materials in the Fashion and Luxury Sector

Since the dawn of civilization, human beings have always felt the need to produce fabrics. In fact, the history of fiber has its roots in the twenty-seventh century BC, with the production of silk (Muthu, 2017). Over centuries, the function of textiles as primary human need has given way to an intensive and massive production. It is testified by a long history of exploitation of natural and animal resources and by the subsequent transition to synthetic fibers, which have led to an unsustainable consumption rate attributed to the fashion industry. Although pollution is often associated with petroleum-based synthetic fibers, such as acrylic, polyester, nylon, and spandex, the environmental impact of resource-intensive natural fibers should not be forgotten. For example, cotton is considered by many eco-friendly; however, its production involves the use of large volumes of chemicals to meet the growing demand, with devastating consequences on the areas where it is cultivated. This is the case of Lake Aral in Central Asia, which has gone from being a rich and fertile land to an arid and almost desert territory because of the excessive farming of cotton (Wood, 2019). These facts depict an undoubtable critical framework for the fashion industry. The hope is to move toward a new era, looking at nature as a source of new eco-sustainable materials. Among the fashion-luxury world, there are already innovative realities that have wisely exploited the inspiration from nature to create innovative organic fibers and sustainable businesses.

The disruptive business model of the Italian company *Orange Fiber* perfectly reflects this statement and provides the perfect blending between the concepts of insular economy and reconstructive sustainability. In Italy more than one million tons of citrus byproducts (orange and lemon peels) are wasted every year (Todeschini et al., 2017). Orange Fiber was created in Sicily in 2014 from the idea of two women, Adriana Santonocito and Enrica Arena, who cleverly understood the potential of using this waste, called *pastazzo*, as raw material for textile production. The mission of the company is "to create sustainable and innovative materials for fashion starting from citrus juice by-products, (…), helping to shape a new concept of luxury based on ethical and sustainable lifestyle" (Orange Fiber Website). The production process, patented since 2014, relies on the extraction of high-quality cellulose from the citrus juice industry scraps, which today accounts for 60% of the original weight of the processed fruits. The peels are processed to extract the cellulose that is spun to form the final yarn, which is then turned into the perfect fiber for the fashion brands worldwide. This green process has the potential to transform the 700,000 tons of *pastazzo* produced annually by the Italian citrus processing industry into high-quality fabrics: black-and-white lace silk, cream-colored satin, and viscose-like fabric. The sustainable drivers of zero waste, local sourcing, recycling, and zero waste are all included into and the Orange Fiber business model. Orange Fiber is born within the Italian citrus district; however in Italy there are two other important districts, which can be involved in the insular theory because of their territorial conformation. It is from this territorial richness that the experiences of *AppleSkin* and *Vegea* are born. Founded in the Italian apple district in Bolzano, *AppleSkin* is a fabric patented by Frumat, and it is a vegan leather-like material made from waste from the apple industry. *Vegea* is a company founded in Milan in 2016 that produces vegan leather from industrial waste from wine production (Noia, 2022).

New textile fibers produced with food waste are manufactured also in Japan, with the *Soybean* Protein Fiber produced from soybean processing waste and with the *Craybon*, from the extraction of chitosan from the exoskeleton of the crustaceans and then mixed with linen or cotton fibers. Meanwhile, in Taiwan a new fiber made from the processing of coffee grounds is produced under the name *S. Cafè* (Noia, 2022).

A different approach to repurpose agricultural waste into low-impact and natural textiles is implemented at Piñatex, company founded by Carmen Hijosa, who, after working as a consultant in the leather goods sector in the Philippines, found in pineapple leaves a sustainable alternative to animal leather. She was inspired by the abundance of natural resources in the Philippines and the use of plant fibers in traditional weaving such as the Barong Tagalog garments. After pineapple harvest, during a process called *decortication*, the leaves are stripped of excess surface biomatter leaving the fibrous inner (Lo, 2006). After the extraction, the fibers go through the washing and drying phase, under the sun or in drying ovens. Once dried, the fibers are subjected to a purification process, and, thanks to the mixture with a corn-based polylactic acid (PLA), after a mechanical process, it results in a non-woven textile, like felt, called *Piñafelt*. This fabric then undergoes chemical treatments through well-established textile finishing techniques, resulting in a material sustainable alternative to leather, which could be implemented across fashion, accessories, and upholstery sectors (Ananas Anam Website). Hugo Boss and Hilton Hotel Bankside are only two of the 1000 brands worldwide who have chosen Piñatex.

The previous cases show how reflection on nature and the surrounding territory can be the inspiration for the creation of new and highly innovative materials. However, innovation in the fashion sector can also go through the rediscovery of eco-friendly natural fibers already known and the improvement of their production process. Among the most environmentally friendly fibers around the world, there is also the eucalyptus fiber. It is a textile with distinctive qualities: it has greater absorbency than cotton, and it's softer than silk and fresher than linen. This material is obtained by reducing the eucalyptus wood in pulp, which is then pressed and spun. Moreover, the fibers from the leaves and stems follow a treatment of boiling, cooling, and spinning. This process is highly environmentally friendly, as trees are grown sustainably in wooded areas that are not suitable for agriculture, nor for the cultivation of any other type of trees. In addition, the water and solvents used in the process are not wasted but can be reused over and over again. It's a closed-loop system, where only recycled and reused materials and no-harmful chemicals are involved (Özlem, n.d.).

The future of the fashion-luxury industry also should pass through the enhancement of another natural textile material, the *bamboo fiber*. Bamboo is one of the fastest-growing plants in the world, whose production is concentered not only in China and India but also in the insular regions of Indonesia. As a fiber, bamboo is a bast lingo-cellulosic fiber (such as jute, flax, ramie, etc.), which is biodegradable and eco-friendly, with antibacterial and UV-protective properties. Its possible usages are not only limited to conventional textiles, but it also fits for high-performance end uses, thanks to its high strength. The highest quality of bamboo fabric, sometimes called *bamboo linen* because of its silky texture, is made with production practices that do not extract cellulose. Instead, it is the end-product of a process, which uses a natural enzyme on crushed bamboo wood fibers. These fibers are then washed and spun into the yarn. This process, defined as *mechanical*, is not harmful to the environment, and the bamboo fabric is a resistant and high-quality material. However, most types of bamboo fabrics are made by the *chemical* process that is unsustainable and incompatible with luxury textile products (Textile Spheres, 23 Oct 2023).

The above-listed fibers are just a few of the possible frontiers that the fashion and luxury industry should address, but there are plenty of other examples of new alternative textiles. One of them is the *hemp fiber*, considered a versatile, antibacterial, durable, and flexible natural material. Its production from hemp, a fast-growing plant which can survive with very little water, is highly sustainable because it doesn't involve herbicides, pesticides, synthetic fertilizers, or GMO seeds. Moreover, an additional environmental benefit is that hemp cultivation releases nutrients into the land, improving soil health rather than harming it. In the fashion industry, many mainstream brands, such as Adidas and Patagonia, have understood the qualities of this raw material, adding it into their items' production and trying to encourage its use as clothing fiber (Preuss, 2017). Meanwhile, new startups are going in the direction of reconstructive sustainability. One example is Hamper, a company that was born from the discovery of a small community of hemp artisans in Nepal, and it is focused on a production where hemp is the protagonist among other natural fibers. Their aim is to build an innovative ecosystem around distinctive artisanal fabrics, natural dyes, and local producers, trying to reincorporate fibers and techniques that are hardly used in the industry (Hemper website).

The lesson that we should draw from these cases is to focus on research and technological development, as well as the exploitation of expertise that is rooted in the local level and the strength in territorial collaboration.

1.3.2 Insular Economy and Ecotourism: A New Paradigm for the Hospitality Industry

Islands are living organisms, where waste management, agriculture, fisheries, tourism, energy, transportation, and health are part of a larger interconnected system. The sustainable development of these ecosystems cannot be realized without a deep understanding of the interdependence between human beings and nature. The ocean itself, while frequently overlooked, holds a significant potential for driving economic growth and sustainability. Shifting toward a blue economy model would foster innovation in sectors like aquaculture, renewable marine energy, biotechnology, and ocean-related endeavors, all of which could have a positive influence on tourism (Saavedra & Alleng, 2020).

The concept of insular economy invokes a resilient and dynamic model that leverages the challenges of the local environment by harnessing and enhancing the available natural resources. Indeed, environmental resources are the *raw materials* for tourism in insular economies, and, especially in small islands, tourism is a vital economic driver. In fact, it accounts for over 25% of the GDP in a number of SIDS (Small Islands Developing States) and makes up 9% of the overall exports (US$61 billion) (World Tourism Organization, 2020). Moreover, island systems are fragile environments that can be easily disturbed by external economic phenomena. Island countries such as Indonesia, Malaysia, and the Philippines are relatively more susceptible to natural hazards, such as tsunamis, hurricanes, global warming, rising sea levels, and related risks due to their geographical isolation (Ebi et al., 2006). Thus, it's crucial to assure the conservation and the sustainable use of these resources to fully benefit from them (Cavallaro & Ciraolo, 2002). Indeed, the well-being of its residents and their economic activities depend on the integrity or degradation of this heritage. The relationship between tourism and the environment is deeply interconnected. It is evident the primary importance of the natural element and its protection for the development of tourism activities. But it is equally necessary to monitor the impact that tourism has on the natural landscape (Lanquar, 1995). The steady increase in tourism may jeopardize the delicate local carrying capacity and cause noticeable instability in the environmental equilibrium. If the tourism sector were to be regulated sensibly, it would undoubtedly present a chance for the island systems to grow sustainably. Considering this, it is seen to be vitally important to rethink a development model that, by utilizing the right instruments, encourages ecotourism while also bringing back to life other long-abandoned productive sectors (Cavallaro & Ciraolo, 2002). The recognition and enhancement of certain agricultural and artisan's production would allow the genetic maintenance of the natural landscape and encourage the development of alternative forms of tourism, leading to the prolongation of the tourist seasons and the creation of new services and thus new jobs for the local population (Gallia & Tre, 2018).

The concept of *ecotourism* combines ethical travel with sustainable tourism, reducing activities that affect the environment, involving the local communities, and preserving the environmental biodiversity and the cultural heritage (Hsiao et al., 2021). Instead of aiming for fast revenues and rapid development at the expense of the lifestyle of local communities, ecotourism promotes initiatives for the environmental conservation and for the support of local sustainable entrepreneurial realities. A concrete example of ecotourism comes from the Galápagos Islands. The Ecuadorian government and the natives of the islands have worked together to promote the conservation of flora and fauna. For example, regulations have been introduced for the protection of the Galapagos National Park Directorate (GNDP) and to promote the products of local residents. In addition, local tourist companies have introduced a waste recycling station to minimize the pollution caused by tourists. Some other companies have worked to incorporate hybrid boats that move thanks to solar energy in order to minimize the use of oil and gasoline near the coasts (Harvard International Review, 2022).

The world of luxury hospitality has already moved in the direction of ecotourism. In particular, there are some sustainable luxury resorts in the Indian Ocean islands that are worth considering. Heritage Le Telfair Golf & Wellness Resort in Mauritius, which has been awarded the Green Key eco-label, and the impressive Oberoi Mauritius Resort provide visitors with carbon-neutral stays with carbon captured through local reforestation programs. Moreover, the resort distributes surplus food to those in need, recruits people within its local community, and uses only regional and organic ingredients in its menu. Another example comes from Seychelles, the Hilton Seychelles Labriz Resort. It is nestled within the Silhouette Islands, whose 85% of its surface is a natural reserve. The resort has tackled plastic pollution from eradicating single-use bottles by installing a water bottling plant on Mount Dauban and providing filtered water into reusable glass bottles. Moreover, it has partnered with the International Fund for Agricultural Development, using almost only (80%) vegetables cultivated locally. All food waste is converted into fertilizer, used by organic farms and native plant nurseries. This is also one of the key strategies at the Raffles on Praslin Island. Environmental and marine protection is also supported by Amilla Maldives Resort & Residences, which is focused on reef conservation and works to support the regeneration of local coral reefs through coral frame planting. Moreover, the resort grows fresh produce in its gardens and uses reusable recycled products made from ocean-bound plastic (Winged Boots, 2023).

The hope for the future of the luxury and fashion industry is to understand and metabolize the principles of ecotourism: environmental defense, preservation of local cultural heritage, community involvement, economic gain, and vulnerable groups’ empowerment.

2 Conclusion

In this chapter, we have analyzed the evolution of the circular economy, which unfortunately shifted from its original objectives, leading to greenwashing rather than tangible, lasting change.

However, insular economies provide a fresh perspective on sustainability, illustrating how constraints can be transformed into strengths through resourcefulness and sustainable practices. Case studies like the *obra de llatra* in Majorca and Casa Flor Ixcaco in Guatemala highlight the benefits or artisanal and local approaches, aligning with the principles of “slow fashion.” Craftsmanship not only prioritizes quality and environmental preservation but also supports local economies and tradition. The challenge ahead is to make craftsmanship accessible to all, thereby fostering a sustainable future for everyone.

In the midst of our current complex environmental crisis, insular economies shine as examples of regenerative systems. They focus on establishing synergies between humanity and nature, following a “recover, restore, and regenerate” paradigm. The fashion industry can learn valuable lessons from these insular realities, where permanent challenges are turned into strengths. We’ve also explored innovative startups like Orange Fiber and Piñatex, which draw inspiration from the nature to create organic fibers. However, true innovation in the fashion sector should also promote the rediscovery and enhancement of natural fibers such as eucalyptus, hemp, and bamboo.

Environmental resources also play a pivotal role in tourism. In accordance with insular economy principles, ecotourism initiatives aim to conserve these resources and support local communities. We have discussed successful examples within the luxury hospitality sector that are embracing the principles of ecotourism and reconstructive sustainability. These cases underscore the importance of investing in the protection of natural and cultural heritage, local resources, community development, and territorial collaboration.

Ultimately, the “insular economy” concept extends beyond the geographical confines of islands, offering a resilient model applicable to areas worldwide facing geographical challenges. It centers on the utilization of local resources and the revitalization of local crafts, providing a sustainable path forward for various regions across the globe.

Bibliography

Part I

Agro, C. (2018). How fast fashion adds to the world's clothing waste problem (marketplace). *CBC News*. Available at: https://www.youtube.com/watch?v=elU32XNj8PM. [9 min 30 sec; 12 min 10 sec]. Accessed 25 Apr 2023.

Cabanes, B. & Cruaud, N. (2023). Construction, textiles: How eco-design is transforming certain markets. *Polytechnique Insights*. Available at: https://www.polytechnique-insights.com/en/columns/industry/construction-textiles-how-eco-design-is-transforming-certain-markets/

Centobelli, P., Abbate, S., Nadeem, S., & Garza-Reyes, J. (2022). *Slowing the fast fashion industry: An all-round perspective*. Elsevier.

Cheng, V. (2021) How has our consumer behavior fuelled the throwaway culture of fast fashion? *Koup*. Available at: https://www.koup.co/blogs/the-koup-blog/the-throwaway-culture-of-fast-fashion

Coulson-Drasner, A. (2021). H&M and Zara: Can fast fashion be eco-friendly? *DW Planet A*. Available at: https://www.youtube.com/watch?v=00NIQgQE_d4.

DiSalvo, M. (2019). Can recycled cotton make fast fashion sustainable? *Virtue+Vice*. Available at: https://shopvirtueandvice.com/blogs/news/recycled-cotton-fast-fashion-sustainable

Domingo. (2021). Économie circulaire: Pourquoi il faut se méfier des solutions de recyclage trop « évidentes ». *20 Minutes × The Conversation*. Available at: https://www.20minutes.fr/planete/3134035-20210929-economie-circulaire-pourquoi-faut-mefier-solutions-recyclage-trop-evidentes

Domingo. (2023) Eco-design should be favoured over recycling. *Polytechnique Insights*. Available at: https://www.polytechnique-insights.com/en/columns/society/eco-design-should-be-favoured-over-recycling/

Dontoh & Pandya. (2022). What actually happens to your donated clothes. *Bloomberg Quicktake*. Available at: https://www.youtube.com/watch?v=8Jbqup_MYjc

Ellen MacArthur Foundation. (n.d.). *What is a circular economy?* Circular Economy Introduction. Available at: https://www.ellenmacarthurfoundation.org/topics/circular-economy-introduction/overview

Hepster. (2023). *Le mythe de l'économie circulaire – de quoi s'agit-il concrètement et est-ce vraiment réalisable?* Available at: https://partner.hepster.com/fr-fr/insights/blog/page/circular-economy

Igini, M. (2022). *10 concerning fast fashion waste statistics*. Earth.org. Available at: https://earth.org/statistics-about-fast-fashion-waste/#:~:text=1.,on%20landfill%20sites%20every%20second.

Insider Business. (2022). How 7.5 million pounds of donated clothes end up at a market in Ghana every week | World Wide Waste. *Insider Business*. Available at: https://www.youtube.com/watch?v=Ichj0ac3V_Y

Yeoh, M. (2018). *UNDP goodwill ambassador Michelle Yeoh follows the sustainable fashion trail*. United Nations. Available at: https://youtu.be/Ko8qCyoZ-Mg

Part 2

Artisans fair trade Guatemala/Santiago Atitlan. El Pelicano. Available online: https://www.elpelicano-store.com/commerce-equitable/

Casa Flor Ixcaco. Woven-GT. Available online: https://www.woven-gt.com/

Culture and identity: Indigenous clothing in Guatemala. Mil Milagros. Available online: https://www.milmilagros.org/story/indigenous-clothing-in-guatemala

Deidda, M. (2014). *Insularity and economic development: A survey*. Centro Ricerche Economiche Nord Sud (CRENOS). Available online: https://crenos.unica.it/crenos/sites/default/files/WP14-07_0.pdf

Donaldson, T. (2020). Is slow fashion the new luxury? *WWD (Women's Wear Daily)*. Published on September 15, 2020. Available online: https://wwd.com/feature/slow-fashion-artisan-luxury-sustainability-social-impact-1234575062/

How Mayan weavers are keeping a 2,000-year-old indigenous craft alive. *Business Insider Today*. Published on August 5, 2021. Facebook video link: https://www.facebook.com/BusinessInsiderToday/videos/how-mayan-weavers-are-keeping-a-2000-year-old-indigenous-craft-alive/2189915994648845/

Islands as model environments. *Journal of Island and Coastal Archaeology*. Available online: https://www.researchgate.net/publication/324756562_Islands_as_Model_Environments

Lieberman, G. (2019). *2030 global consumer trends*. Director of Trends and Social Media Research, America. Published in November 2019. Available online: https://canipec.org.mx/wp-content/uploads/2020/03/2030-global-consumer-trends.pdf

New and good era in fashion: Artisans. HechoXNosotros. Published on June 17, 2021. Available online: https://www.hechoxnosotros.org/post/new-and-good-era-in-fashion-artisans

Piper, A., Townsend, K., & Jabur, L. (2023). Resilience, resourcefulness and creativity: Learning from the diversification of guatemalan artisans during the pandemic to sustain textile traditions. *Journal of Textile and Fashion Sustainable Studies*. Published online on June 19, 2023. Available online: https://www.tandfonline.com/doi/full/10.1080/20511787.2023.2212515

See how indigenous weaving styles are preserved in Guatemala. National Geographic. Available online on Youtube: https://www.youtube.com/watch?v=apLl5H1uslM

Spilannis, I., Kizos, T., Koulouri, M., Kondylia, J., Vakoufaris, H., & Gatsis, I. (2008). *Monitoring sustainability in insular areas*. Available online: https://www.researchgate.net/publication/222398220_Monitoring_sustainability_in_insular_areas

Sustainability in fashion: Why artisanal support is important. Matter Prints. Available online: https://www.matterprints.com/journal/community/why-artisanal-support-is-important/

The impact of sustainable fashion on local communities and Artisans. Real Thread. Available online: https://www.realthread.com/blog/the-impact-of-sustainable-fashion-on-local-communities-and-artisans

The Mayan textile tradition: Towards a sustainable fashion. Published on March 01, 2018. Trama Textiles Blog. Available online: https://tramatextiles.org/blogs/trama-blog/the-mayan-textile-tradition-towards-a-sustainable-fashion

UNESCO. *Traditional craftsmanship – Intangible heritage – Culture sector*. Available online: Savoir-faire liés à l'artisanat traditionnel – patrimoine immatériel – Secteur de la culture – UNESCO

Part 3

Ananas Anam Website. https://www.ananas-anam.com/about-us/

Bloomberg. (2022). The global glut of clothing is an environmental crisis. https://www.bloomberg.com/graphics/2022-fashion-industry-environmental-impact/

Cavallaro, F., & Ciraolo, L. (2002). *Economic and environmental sustainability: A dynamic approach in insular systems*. Available at SSRN 307199.

Ebi, K. L., Lewis, N. D., & Corvalan, C. (2006). Climate variability and change and their potential health effects in small Island states: Information for adaptation planning in the health sector. *Environmental Health Perspectives, 114*(12), 1957–1963.

Gallia, A., & Tre, R. (2018). *Il paesaggio rurale insulare. L'isola di Ponza tra sedimentazioni storiche e processi di valorizzazione* (pp. 195–207). Tre Press.

Harvard International Review. (2022, October 14). The Galápagos Islands: Economy over ecosystems. https://hir.harvard.edu/the-galapagos-islands-economy-over-ecosystems/

Hemper website. https://hemperstore.com/pages/historia

Hsiao, C. Y., Kuo, C. M., & Tuan, C. L. (2021). Island ecological tourism: Constructing indicators of the tourist service system in the penghu national scenic area. *Frontiers in Ecology and Evolution, 9*, 708344.

Lanquar, R. (1995). *Tourisme et environnement en Méditerranée: enjeux et prospective*.

Lo, J. (2006). Design, manufacturing and applications of composites. In *Proceedings of the Sixth Joint Canada-Japan Workshop on Composites*. DEStech Publications, Inc.

Muthu, S. S. (Ed.). (2017). *Sustainable fibres and textiles*. Woodhead Publishing.

Noia, E. (2022). *Materiali tessili da scarti alimentari: rifiuti diventano risorsa*.

Orange Fiber website. https://orangefiber.it/

Özlem, K. A. Y. A. (2021). *Innovative fabrics and products that can create a revolution in fashion*. XV. International Turkic Culture, Art and Protection of Cultural Heritage Symposium/Art Activity, Turkey.

Preuss, S. (2017). *Sustainable textile innovations: Hemp fibers*. https://fashionunited.uk/news/business/sustainable-textile-innovations-hemp-fibres/2017071025112

Saavedra, J. J., & Alleng, G. P. (2020). *Sustainable Islands: Defining a sustainable development framework tailored to the needs of islands* (IDB Tech. Note, 2053, pp. 1–72). IDB.

Sun, Y. (2023). Sustainable luxury circular and economy. In *Managing luxury brands: A complete guide to contemporary luxury brand strategies* (p. 109). Eleonara Cattaneo, Kogan Page.

Textile spheres. (2023, October 23). https://www.textilesphere.com/2020/09/bamboo-fiber.html

Todeschini, B. V., Cortimiglia, M. N., Callegaro-de-Menezes, D., & Ghezzi, A. (2017). Innovative and sustainable business models in the fashion industry: Entrepreneurial drivers, opportunities, and challenges. *Business Horizons, 60*(6), 759–770.

Winged Boots. (2023, April 17). *Six of the best sustainable hotels in the Indian Ocean*. https://www.wingedboots.co.uk/magazine/six-of-the-best-sustainable-hotels-in-the-indian-ocean/

Wood, J. (2019). Bioinspiration in fashion – A review. *Biomimetics, 4*(1), 16.

World Tourism Organization. (2020). *UNWTO briefing note – Tourism and COVID-19, Issue 2. Tourism in SIDS – The challenge of sustaining livelihoods in times of COVID-19*. UNWTO. https://doi.org/10.18111/9789284421916

Index

S. S. Muthu (ed.), *Corporate Social Responsibility in Textiles and Fashion*, Sustainable Textiles: Production, Processing, Manufacturing & Chemistry, https://doi.org/10.1007/978-3-031-61099-8

The manufacturer's authorised representative in the EU is Springer Nature Customer Service Centre GmbH, Europaplatz 3, 69115 Heidelberg, Germany. If you have any concerns regarding our products, please contact ProductSafety@springernature.com

Printed and bound by CPI Group (UK) Ltd, Croydon, CR0 4YY
30/09/2026
02225563-0004